FEMINIST TECHNICAL COMMUNICATION

FEMINIST TECHNICAL COMMUNICATION

Apparent Feminisms, Slow Crisis, and the Deepwater Horizon Disaster

ERIN CLARK

UTAH STATE UNIVERSITY PRESS
Logan

© 2023 by University Press of Colorado

Published by Utah State University Press
An imprint of University Press of Colorado
1580 North Logan Street, Suite 660
PMB 39883
Denver, Colorado 80203-1942

All rights reserved

   The University Press of Colorado is a proud member of the Association of University Presses.

The University Press of Colorado is a cooperative publishing enterprise supported, in part, by Adams State University, Colorado State University, Fort Lewis College, Metropolitan State University of Denver, University of Alaska Fairbanks, University of Colorado, University of Denver, University of Northern Colorado, University of Wyoming, Utah State University, and Western Colorado University.

∞ This paper meets the requirements of the ANSI/NISO Z39.48-1992 (Permanence of Paper).

ISBN: 978-1-64642-526-6 (hardcover)
ISBN: 978-1-64642-527-3 (paperback)
ISBN: 978-1-64642-528-0 (ebook)
https://doi.org/10.7330/9781646425280

Library of Congress Cataloging-in-Publication Data

Names: Clark, Erin (Erin A.), author.
Title: Feminist technical communication : apparent feminisms, slow crisis, and the Deepwater Horizon disaster / Erin Clark.
Description: Logan : Utah State University Press, [2023] | Includes bibliographical references and index.
Identifiers: LCCN 2023023503 (print) | LCCN 2023023504 (ebook) | ISBN 9781646425266 (hardcover) | ISBN 9781646425273 (paperback) | ISBN 9781646425280 (ebook)
Subjects: LCSH: Feminism and rhetoric. | Feminist theory—Social aspects. | Communication of technical information—Social aspects. | Technical writing—Social aspects. | Risk communication—Social aspects.
Classification: LCC HQ1176 .C53 2023 (print) | LCC HQ1176 (ebook) | DDC 302.2082—dc23/eng/20230609
LC record available at https://lccn.loc.gov/2023023503
LC ebook record available at https://lccn.loc.gov/2023023504

Cover photograph by Sharon Pittaway, https://unsplash.com/@sharonp

For Caroline and Sammy

CONTENTS

List of Illustrations   ix

Acknowledgments   xi

Preface: On Positionality and Inclusion   xiii

1. Feminist Technical Communication   3
2. Apparent Feminisms   35
3. Slow Crisis   68
4. Disaster   85
5. An Apparent Feminist Analysis of the Deepwater Horizon Disaster   106
6. Looking Forward, Looking Back   123

Notes   149
References   155
Index   177
About the Author   181

ILLUSTRATIONS

FIGURES

4.1. Boom in the Gulf   91
4.2. Dauphin Island berm, June 2010   92
5.1. Area closed to fishing as a result of the Deepwater Horizon spill   108

TABLES

0.1. Selected history of voting rights in the United States   xvi
4.1. Selected, localized, and traditional time line of the Deepwater Horizon Disaster   89

ACKNOWLEDGMENTS

I am grateful to the Department of English and the College of Arts and Sciences at East Carolina University for their sponsorship of a National Humanities Center (NHC) residency that supported this work. I am equally grateful to the NHC and particularly to library director Brooke Andrade for their assistance. Doctoral students Bess McCullouch and Morgan Banville were enthusiastic and meticulous copyeditors and citation auditors, and they relied on the generous intellectual contributions of Cana Uluak Itchuaqiyaq to do their work. In addition, my thanks to colleagues and friends Kellie Sharp-Hoskins, Marie Moeller, Angela Haas, Michelle Eble, Nikki Caswell, Andrea Kitta, Helen Dixon, Laura Mazow, and Will Banks for the intellectual community they help sponsor.

This book is a project that was a long time in the making, and as such it includes many bits and pieces that have been previously published. Portions of this book are revised from my dissertation "Theorizing an Apparent Feminism for Technical Communication," completed at Illinois State University under the direction of Angela Haas (Frost 2013a). Other portions of this book are revised from an article published in the *Journal of Business and Technical Communication* (Frost 2016); portions are also drawn from an article published in *Technical Communication Quarterly* (Frost 2013b).

PREFACE

On Positionality and Inclusion

The world needs more feminisms. Technical communication needs more feminisms. I will demonstrate both assertions throughout this book, but first I want to focus on the plurality of feminisms and to account for my own approach and how it frames this book. My 2013 dissertation project was born in part out of the difficulties of dealing with a term that has so many deeply felt meanings for so many people. In fact, my interest in feminisms has as much to do with the term's rhetorical velocity (Ridolfo and Devoss 2009) as with its history. At the outset, it's important to be clear that this book understands feminisms' ultimate importance as a given but simultaneously wrestles with their contextual relevance and meaning. A brilliant mentor once suggested that we tend to study what we're bad at. In my case at least, I think she's right; the multiplicity of meaning evoked when we discuss feminism is something I've always struggled to get a handle on. When someone says they are a feminist or when someone says they are *not* a feminist, what do they mean? How can any sort of coherent movement or social agenda exist behind a term that is so fractured, so individual? If at times it seems that I am either disavowing or too vigorously defending feminism, it is because of a deep attachment to the thing that term evokes—the notion of equity among the sexes—and a realization that we are not very close to that goal. As far as the type of feminist engagement I offer here, then, I can say this: I attempt to privilege this notion of sexual equity and to forward a rhetorical and intersectional brand of feminism, and I believe apparent feminism is a framework for doing these things that has a lot to contribute to and beyond technical communication as a discipline.

While I have built this methodological approach using bits and pieces of work from scholars of a variety of backgrounds, including especially differing ethnicities and nationalities as an attempt to check my own privileges, I want to acknowledge at the outset that I—the assembler of this particular arrangement of theories—am white. That information is important at this cultural moment, when state-sanctioned violence against Black people continues and technical communication as a field has recognized some of the impulses that undergird this reality within itself. Thus, transparency of positionality and reflexivity is critically important for technical communication scholars who hope to engage in social justice work at this moment. I believe my explicit acknowledgment of my embodiment is important here so that readers can read this work through the lens of that embodiment. As Andrea Olinger said in her closing remarks for the 2021 Watson Conference: "When we do something that causes harm, we . . . will take responsibility for the harm. We deliberately say 'when,' not 'if,' we do harm. Because we participate in a system founded on and fueled by white supremacy, failing our marginalized colleagues, however good our intentions, is a tragic given."

Apparent feminism as a theory was only ever meant to be temporary and permeable, and that is precisely because I cannot imagine or comprehend every positionality; no one can create a one-size-fits-all feminism. Because I am white, my particular brand of feminism may have—in fact, certainly *does* have, though I have striven to eliminate them—weaknesses or peculiarities that are particular to my view of the world in terms of race privilege. While some scholars of color have used apparent feminism productively in their own work, it must be said that I could never have leveraged this very same methodology in the ways they do—and their uptake of this methodology in ways beyond what I had previously imagined is most welcome.

As part of my attention to positionality and inclusivity, I strive to be specific in my terminology when it comes to identity. For example, when I later discuss pregnancy and related rhetorical situations, in some places I will refer to pregnant people and in some places I will refer to women. My rhetoric operates according to the following logics/facts: women are women. I do not distinguish between transwomen and ciswomen unless there is a specific, experience-based reason to do so, which there rarely is. When I refer to pregnant people, I mean people who are pregnant, whether they identify as transmen, ciswomen, genderqueer, or nonbinary. Finally, women (in addition to pregnant people) are an identity group whose right to privacy has been generally diminished as a result of the medico-legal establishment's treatment of pregnancy *regardless of an individual person's actual ability to become pregnant*. That is, transmen and infertile women—anyone "read" as a woman—are also impacted by social, medical, and legal transgressions vis-à-vis bodily autonomy. As one reviewer of this book smartly put it, "If it's traumatizing for cis women, it is for trans men as well."

This care for terminology is important. Rhetoricians, technical communicators, compositionists, feminists, and many other readers can probably agree at least on that much.[1] Temptaous Mckoy (in Mckoy, Shelton, Davis, and Frost 2022, 75) argues that the phrase *white men* sometimes serves "as a surrogate for institutionalized patriarchal systems, instead of forcing us to deal with the more complex notion that patriarchy happens as a result of male dominance but is perpetuated by all kinds of people." Mckoy points out that the sort of thinking that permits the term *white men* to stand in for patriarchy is the precise sort of thinking that gives rise to white feminism. The assumption of a monolith in terms of identity almost never serves well as a reflection of actual people.

As others have, I argue that white feminism—or feminism that understands only white women as its stakeholders without marking itself as such—is not feminism at all. Equity among the sexes necessarily means equity across all identity markers as well. The ratification of the Nineteenth Amendment to the US Constitution provides a useful example (table 0.1). Celebrating the anniversary of this historic event, as we did a few years ago, is a fine thing, but our framing of such a celebration is critical. If we celebrate it as the anniversary of "women's suffrage," we erase the women who did not gain suffrage with this legislative change; as we know, various forms of technical communication have historically been used to disenfranchise Black voters in particular (Jones and Williams 2018). But if we celebrate it as the anniversary of "white women's suffrage," we are better able to acknowledge a major victory for those striving for sexual equity that nevertheless has a history heavily fraught with racism.

I use the term *feminism* to describe what I do in part because it is a body of work I feel authorized to speak within. This does not discount many worthwhile parallel projects; in fact, apparent feminism seeks to find or sustain avenues for conversations among those projects. To better explain: despite a history deeply implicated with racism in this country,[2] feminism has always been a collaborative project. More specifically, people of color have always been part of feminist work, since its inception—though they may not always have called their work feminist, and for good reason. As Ruby Hamad (2020, xv) states, "Writing about race is a fraught business, as is writing about gender. Words and phrases you assume would be easily received exactly as you intended them are bafflingly interpreted as something else entirely." In the face of

Table 0.1. Selected history of voting rights in the United States

Legislative Action	Legislative Vehicle	Effect / Context	Year Ratified/ Enacted
Electoral College founded / voting begins	Article II of the US Constitution (with significant support from implicit knowledge)	White male property holders can vote	1788
Non-property owners granted the right to vote	Various state laws (New Hampshire was the first to eliminate property holding as a requirement; North Carolina was last.)	This suffrage expansion benefits only white men.	1792–1856
The right to vote cannot be denied based on "race, color, or previous condition of servitude."	Fifteenth Amendment	Jim Crow laws (e.g., poll taxes, literacy tests, records requirements) widely disenfranchise Black men.	1870
The right to vote cannot be denied "on account of sex."	Nineteenth Amendment	White women's suffrage realized in most contexts. Jim Crow laws (e.g., poll taxes, literacy tests, records requirements) widely disenfranchise Black women.	1920
Poll taxes eliminated	Twenty-fourth Amendment	Black suffrage realized in most contexts.	1964
Discriminatory practices like literacy tests outlawed	Voting Rights Act of 1965		1965
Voter ID required in about two-thirds of states	Various state laws	Continuing debate over the ethics and effects of voter ID laws	

such interpretation, particularly when that (re)interpretation happens among those with power and privilege, it's sometimes better to move out from under the onus of a loaded term entirely. Womanist theory emerged in the late 1970s largely in response to white feminism (Walker 1983). Womanism has been theorized as a broader category than feminism and one that provides space for Black women to prioritize their Blackness in ways both white feminisms and feminisms writ large had not allowed/do not allow. Black feminisms have additionally provided a powerful counterpoint to white feminisms (Hill Collins 1990), and attention to intersectionality (Crenshaw 1991) has become perhaps the most important marker of progressive feminisms. Ideally, womanism, Black feminisms, and feminisms all desire sexual equity for/among humankind, though they may sometimes differ in their sense of the current state of progress and in methods toward achieving said equity.

Some feminisms also have an unfortunate history of mirroring sexist gatekeeping in their treatment of gender. To back up a bit, we know that sexism is the culturally ingrained notion that a person is superior because of their sex and that generally we (not just men; all of us) tend to privilege men. This has been shown empirically across many, many contexts and is widely evidenced anecdotally. One of my favorite anecdotes demonstrating egregious-to-the-point-of-funny sexism in the sciences is the case of Ben Barres, a transman, who "overheard another scientist say, 'Ben Barres gave a great seminar today, but his work is much better than his sister's work'" after his transition; the scientist was, of course, referring to the very same person in the comparison (Stanford University Medical Center 2006, para. 4). The only difference was the way the speaker perceived the sex of the person whose work he was judging. Unfortunately, some feminists—and thus some feminisms—also practice this kind of sex-based exclusion, and in more conscious ways. A popular example of this is the (in)famous and now defunct Michigan Womyn's Music Festival, which backed a women-born-women policy and barred transwomen from attending. These types of actions gave rise to the term TERFs—trans-exclusionary radical feminists—and sparked conversations about diversity and inclusion within feminist circles that continue and remain needed to this day.

I could write many chapters about why our artificially dimorphic sexual organizing system is a bit odd in the face of nature to begin with and many more chapters about how interesting the impulse to organize ourselves and everything else according to chromosomes, hormones, or anatomy is. (These things have been written already, if not widely taken up or acted upon.) We know, of course, that being trans relies on buying into this system of sexual dimorphism (Serano 2013)—which we all do, at some level, to abide by the social contract and thus earn our place in society. Ultimately, though, we also know that the allure of neat categories is false and that essentialism begets exclusion. As Chandra Talpade Mohanty (1988, 2003) has eloquently argued for many years, to reinforce a monolithic understanding of a group is to do violence to that group and to risk misunderstanding it entirely. Mohanty critiques Western feminists' use of the term/idea of *third world women* and in the same way shows that we cannot boil "women" down to a specific essence. Another way of thinking about this is that there is no single common experience of being a woman; in fact, my experiences of identity formation growing up in the midwestern United States may be more similar to a man's growing up in a similar context than to those of a woman from somewhere in the Global South. It is true that a man who grows up similarly to me is still socialized differently than I am, but who am I to judge which particular differences matter and to what degree?

Despite feminists' shortcomings and complications—and because of them—the world needs feminisms. More specifically, we need apparent feminism because of its explicit modeling of inclusivity, its built-in reflexivity, and its constructive criticism of some of technical communication's most valued terms. (The latter, at least, requires a certain amount of indiscipline ethos, privilege, or both to accomplish.) Apparent feminism as a methodology is not perfect, and neither is it perfect as a practice. I will undoubtedly write things that require correction, calling in, and/or calling out. But I'd rather do the hard work of trying to contribute usefully to the important conversations we—feminists and allies, technical communicators—are having about gender, race, ability, orientation, and *identity* than sit by and implicitly ask others to do that labor for me. Floyd Pouncil and Nick Sanders (2022) describe a Black–feminist-oriented model for coalitional relationships that I find helpful in conceptualizing the trajectory of the work of inclusion. In their model, a key element is *upward critical collaboration* that happens in an iterative process with both *inward* and *outward critical reflection*. Inward critical reflection, they say, involves questioning one's own positionality. Outward critical reflection asks us to position ourselves in relationship to others with a shared identity and to "recognize multiplicity in a shared identity" (287). We might then return to inward critical reflection, or we might move toward upward critical collaboration, in which we form coalitions with others who are unlike ourselves but who share a goal or commitment. Pouncil and Sanders note that, in practice, we are usually doing all three at once.[3]

Another element of this type of engagement is that people with power and privilege—I am counting myself here, in this context, as a white associate professor—have to be willing to be uncomfortable. Emily January Petersen and Rebecca Walton (2018) articulate this well as one of their nine recommendations for feminist work in technical communication. They also recommend that we recognize that feminisms are intrinsically tied to social justice work and to decolonial work, acknowledge (our complicity in) existing systems of domination and oppression, interrogate claims of universal knowledge, and challenge dominant perspectives. Engaging and championing intersectional feminisms in this way is hard work. This work takes a lot of deliberate rhetorical/mental labor and a lot of unlearning, the latter particularly for those of us with racial privilege. But at the same time that it's hard work, it's also not much to be asked to be responsible for, in the scheme of things.

I acknowledge, perhaps conversely, that writing books is rarely the most important work a feminist could do. The value in book writing, at least for me, lies in the different trajectories this genre allows for contributing to the larger conversation—a different audience, a different valence and context for arguments I've made previously. As such, some of what is contained here is not new. The nature of publishing means that nothing much is very new when it finally hits readers' hands, particularly by the standards of technical communication and its relative speed, and that means that some of what we (technical communicators *and* feminists) write is for posterity—for the sake of preserving a conversation we can return to, teach students with, ruminate on. In short, I rather hope that this book is not the most important feminist work I have done or the most important feminist work I will do—though it is a reflection on or distillation of some of that work, conveyed in a way that might allow it to serve as a model of what to do and build from or perhaps in some cases what not to do (though I hope not too often). In my opinion, some of the most important work feminists do is in teaching and amplifying others (Mckoy 2019), which I commit to doing in this book and also elsewhere—particularly women of color, whose "strong objectivity" (Harding 1992, 458) stands to benefit us all, perhaps especially the field of technical communication: "Strong objectivity requires that the subject of knowledge be placed on the same critical, causal plane as the objects of knowledge. Thus strong objectivity requires what we can think of as 'strong reflexivity.' This is because culture-wide (or near culture wide) beliefs function as evidence at every stage in scientific inquiry."

Harding's often misunderstood notion of "strong objectivity" posits that true objectivity[4] is impossible and thus that people who occupy positionalities that have been marginalized are those best situated to pull the curve of bias back toward some sort of balance. Sandra Harding is, of course, working in the sciences, but Cecilia Shelton (2019b) offers us a way of thinking about the value of standpoint perspectives not just in technical communication but also in the technical communication *classroom*—the main site of activism for many of those who identify as technical communication scholars. Alicia Hatcher (2021) similarly demonstrates that some methods of persuasion are not only more effective but also only recognizable as persuasive communication when enacted by people who occupy marginalized positionalities; her concept of performative symbolic resistance offers a way to describe and analyze activist work.

As should be apparent by now, I value feminist standpoint approaches and want to enact and enable them whenever possible. To that end, a recent editorial in *Trends in Cognitive Sciences* proves useful as an activist moment in the context of scholarly publication. Perry Zurn, Danielle S. Bassett, and Nicole C. Rust (2020, 669) advance the prospect of a "citation diversity statement" that is aimed at letting scientist-authors demonstrate how they are helping "science become more inclusive and diverse overall." Their editorial includes a link to a form where authors can check boxes next to statements like "we worked to ensure gender balance in the recruitment of human subjects." Of particular interest is a section toward the end of the form that addresses "helicopter science," or science that relies on members of low-income or Indigenous groups, and asks if authors have made a good-faith effort toward meaningful inclusion. Near the end of the form, authors are offered the suggestion to copy-paste all the statements they checked and include them as a statement in the text of their submitted manuscript. Such a statement might look something like this: "I worked to ensure gender balance in the recruitment of human subjects. I worked to ensure ethnic or other types of diversity in the recruitment of human subjects. I worked to ensure that the study questionnaires were prepared in an inclusive way. One or more of the authors of this paper self-identifies as living with a disability. While citing references scientifically relevant for this work, I also actively worked to promote gender balance in the reference list."

While the apparency of such a statement is important (though limited), the effect of creating the statement is likely the most valuable part of this process. Filling out the form, particularly if they've reviewed it in advance, will help authors to think about their process and of ways they can better their work. For example, as part of my thinking about responsible citation practice and with the help of then PhD students Morgan Banville and Bess McCullouch, I conducted a citation audit of this book using Cana Uluak Itchuaqiyaq's (2021, original emphasis) list of "*self-identified* multiply marginalized and underrepresented scholars (MMU) in technical communication and related fields." This is one type of endeavor a citation diversity statement can help to encourage. Further, the form described above is widely adaptable, and a version of it could be used with ease in the social sciences and humanities (as I demonstrate here).

Within technical communication, scholars have also done some big-picture, action-oriented thinking about inclusion and diversity. A group of prominent scholars has publicized a Google Doc to offer "explicit guidance on anti-racist professional practices in the form of a heuristic for editors, reviewers, and authors involved in academic reviewing" ("Anti-Racist Scholarly Reviewing Practices" 2021).[5] And a subset of that group published "Participatory Coalition Building: Creating an Anti-Racist Scholarly Reviewing Practices Heuristic," which is an extended explanation of the process followed in developing those practices (Cagle et al. 2021).

I take the importance of being apparent about positionality seriously, and I have made every attempt to privilege inclusivity throughout this text. I have no doubt fallen short in places, and I look forward to the opportunities those shortcomings will generate for learning. Ultimately, despite and because of those anticipated flaws, this book forwards an argument that feminist technical communication is vital for the future of the field, and it offers apparent feminism as a specific theoretical contribution for helping to build a feminist project in the field and beyond.

FEMINIST TECHNICAL COMMUNICATION

1
FEMINIST TECHNICAL COMMUNICATION

INTRODUCTION AND EXIGENCE

This book serves as an introduction to feminist technical communication and argues for intersectional feminist approaches as vital for the future of technical communication as a field. Situating feminisms and technical communication in relationship as the focal point of an entire book is a project that has not been previously undertaken, feminisms are more often understood as one tool in a larger toolkit of cultural approaches to the field. This project does not contradict such approaches, but it also centralizes the importance and magnitude of feminist contributions to technical communication. To that end, it takes an intersectional approach to feminist technical communication and offers relevant histories of a variety of feminist works in the field. As several scholars have noted, a surge of work in feminist technical communication took place in the 1990s and interest in the subject then waned; more recently, social justice has become an important organizing principle in the field. This text seeks to revitalize and intersectionalize feminist technical communication as part of that larger social justice project.

I forward two main arguments, the second predicated on the first. While my framing of intersectional feminist approaches as vital for the future of technical communication is the larger-in-scope argument of this book, and while I hope this argument and its attendant literature reviews will be useful particularly in graduate technical communication classrooms, I also use apparent feminisms to argue more specifically that the traditional efficiency model in technical communication is not an effective or a sustainable approach. If our field is to retain *efficiency* as a guiding principle—Kenneth Burke (1945) would call it a "god term"—then our field's understandings of efficiency must change. Because crisis communication and health communication are the sub-fields that tend to most explicitly call on efficiency models, much of my work is sited in those contexts. Context, of course, is critical to rhetorical understandings of technical communication and feminisms.

Because feminisms are based in material experience and feminist theory can never be separated from the corporeal, I utilize a specific model context. This book explores communication about health effects related to the Deepwater Horizon Disaster (DHD). Some people will say no such effects exist and thus there is no problem to be addressed here. However, patterns of communication surrounding the DHD and (the lack of communication about) health effects demonstrate purposeful rhetorical attempts to steer the conversation in directions that are efficient or expedient for those with the most rhetorical power.

Sometimes a problem is not apparent—but that doesn't mean it is unimportant. Sometimes a problem is not (made) apparent precisely because it *is* important. A relevant example that touches on the context at hand and also demonstrates the problems with lack of apparency is made evident in an investigative story by the National Audubon Society: "Even before the BP [British Petroleum] disaster, the Gulf was a region of neglect. We certainly have not treated it like a spot that deserves to be studied, which would have been helpful. Many scientists say it's practically impossible to determine what the state of the Gulf ecosystem is *now* because we didn't know what it was *then*. As John Dindo, senior marine scientist at Alabama's Dauphin Island Sea Lab, puts it, 'Without that baseline data, you are pulling things out of your ass'" (Gessner 2015, para. 4, original emphasis).

In other words, it is not just lack of apparency that is a problem but also the timing of that apparency. Without prior data, we are unable to effectively extrapolate. This is a problem that will be repeated until we learn to understand the efficiencies of data collection differently. A grittier example demonstrates a future lack: "Defenders' Chris Haney's retrospective study of bird deaths concluded that approximately 600,000 to 800,000 were killed by oil from the BP spill, despite the fact that only 6,147 cadavers were collected and counted in the year after. 'Of all the spills I've ever studied, none had as many combinations of factors that have made it harder for a dead bird to actually reach the morgue and be counted,' he'd told me. 'I fear that the science coming out of the spill won't be a match for the size, scope, and volume of the spill itself'" (Gessner 2015, para. 12). General lack of data can be a problem with apparency, and this is the problem I most often faced in the present project.

However, specific lack of data—like absent bird bodies—is also a potential choke point for studies of crisis situations. As in the example above, an apparent feminist analysis of the DHD faces both these problems of lack of apparency—and this is hardly a new problem for environmental work. Donnie Johnson Sackey's (2020) groundbreaking work in regard to the Flint, Michigan, water crisis and including participants in design work asks environmental justice advocates to consider how policy design that is more inclusive of all stakeholders could have prevented the disaster altogether. He notes that reports of a problem with the water in Flint are documented as far as two years in advance of confirmed contamination (39), meaning that lack of apparency resulted in two years of poisoning for Flint residents. Importantly, Sackey has also worked on a National Institutes of Health grant focused in part on studying means to "more effectively utilize modern modes of communication (e.g., social media such as Facebook)" in relation to risk information (McElmurry 2016, para. 1). In short, Sackey's work, which focuses on environmental justice and technical communication, takes problems of apparency as a first space for intervention.

To attend to such problems of apparency and to narrow the scope of this inquiry to something manageable, I begin with health and medical rhetorics. In examining health communication, it is important to be cognizant of the broader contexts in which health conversations operate. Health and medical rhetorics, a burgeoning sub-field of technical communication, often purposefully blur the lines of health and medicine. Healthcare and medicine are terms with different connotations and varying attachments to ethos, with medicine most often the more respected of the two terms by most measures. Generally speaking, this book engages with health and approaches it as the broader concept—a concept that can encompass not just biological phenomena but also ideas about how economic and ecological health can impact stakeholders.

I focus on healthcare communication with the understanding that it functions, in this instance at least, as a repository and reflection of public understandings of risk and health. This is to say: we understand efficiency in a certain way, and that limits what is recognizable to us about a rhetorical situation. In the case of Deepwater Horizon, efficiency affects what is recognizable about environmental disasters, thus also limiting our possible responses to health as it relates to environment. Our efficiency models tell us where to look and what to look at; they thus implicitly also determine what and who we do not take into account. If we understood efficiency differently, we'd develop different areas of concern and thus have different rhetorical options for moving forward. Kim Hensley Owens (2015) discusses the paradigm shifts that undergird progress narratives and decline narratives. In her case, cultural ideas about the relative safety of hospital births versus midwife births condition us to find data to support our beliefs. It took some time after hospital births became the norm for them to become statistically safer, despite popular conceptions to the contrary.

More recently, we have observed a similar logical problem happening with Covid-19 vaccinations among pregnant women. Kate Cray (2021, para. 2) reported that despite ample evidence that the Covid-19 vaccines are safe and beneficial during pregnancy, "only about 25 percent of mothers-to-be have gotten one during their pregnancy. Rates are even lower for Latina and Black expectant mothers, at 22 and 15 percent, respectively, compared with 27 percent of white and 35 percent of Asian expectant moms." Given our cultural expectations that pregnant women should keep their bodies pure and unaltered by things like alcohol and caffeine, vaccine hesitancy in this group is unsurprising. We are conditioned to believe that pregnant women are behaving as "good mothers" by denying themselves things to protect their unborn child. The problem with this particular cultural logic (as my own high-risk obstetrics team told me repeatedly in the time just before Covid made national headlines, while I was debating the benefits of cesarean delivery) is that pregnant people are then prone to miscalculate risk, construing non-intervention as the least risky approach even when the opposite is empirically true. To offer full context for the aforementioned example, women of color (particularly Black women) have a long history of valid reasons—as long as the history of gynecology—not to trust agents of the medical establishment, meaning their vaccine hesitancy is additionally rooted in a parallel set of cultural narratives that advise them to protect themselves and their families from unjust experimentation (Baker 2017; Cray 2021). It is not surprising that the descendants of Betsey (Washington 2006) and Anarcha (Cox et al. 2008), upon whose un-consenting bodies modern gynecology was built, might have reservations about medical directives. Owens (2015) argues that we, as a culture, will not bestow legitimacy on details that do not fit whatever the current narrative is. This, of course, points to the constructed nature of risk and the culturally relativistic nature of ethical systems.

Health communication is a subject increasingly debated in more mainstream arenas in recent years. The Covid-19 pandemic is partly responsible for this, but maternal health is also seeing increased scrutiny. In 2017, tennis phenom Serena Williams suffered a pulmonary embolism (PE) after giving birth. She knew what was happening, as she had had a PE previously, but she had to advocate fiercely for herself to get medical professionals to take the situation seriously. Her story was something of a lightning rod for both the medical community and Black birth-givers. "In 2018 journalists started to tell the stories of people that were dying in childbirth," said Neel Shah, professor of obstetrics and gynecology at Harvard School of Medicine. "Those stories ended up compelling the federal government to start tracking maternal mortality much more systematically" (Eiselt and Lee 2022). Shah's quote is delivered in the recently released documentary *Aftershock*, which follows the families of Shamony Gibson and Amber Rose Isaac after the two women, both Black, died from childbirth complications and medical failings. In the documentary, during a meeting with (Brooklyn) Weeksville Heritage Center's deputy director Anita Warren, Shamony's mother, social worker Shawnee Benton-Gibson, notes that she does reproductive justice work and it didn't make a difference in the outcome for her daughter. "Knowledge doesn't save you," she says.

Benton-Gibson's implicit emphasis on systemic change suggests one way that coalitional feminist models can be useful in community-engaged contexts. One relevant and excellent example of community engagement is when North Carolina Agricultural and Technical State University associate professor Kimberly C. Harper (2022) went on NCImpact to explain what a birth doula is and why doulas are important interventionists for Black women. "Black women are reporting repeatedly that they feel dismissed and their concerns are not heard," Harper said (a statement Shah echoed, speaking specifically in regard to pain). "[A] doula can step in and assist with getting the care that you need . . . Doulas are, I think, integral in changing the maternal landscape in this country." Doulas, of course, function as technical communicators in that they serve as a bridge between different kinds of experts and across various cultures.

The intersection of feminist technical communication and health communication is a rich site to get at cultural relativism. In this book, I use an apparent feminist theoretical lens—with attending ideas about the place of technical communication scholarship and health and medical rhetorics—to demonstrate how reconsidering understandings of efficiency in environmental disaster situations subsequently changes possible approaches to acceptable risk related to healthcare. Researchers in technical communication have addressed the confluence of risk, ethics, and healthcare communication (Ding 2009, 2012, 2013; Faris 2019; Itchuaqiyaq, Edenfield, and Grant-Davie 2021; Lundgren 1994; Youngblood 2012) as well as risk and identity (Chandler and Sano-Franchini 2020; Ruiz 2018). However, rapidly developing technologies and the changing role of media consumers and producers renders this an area where study remains necessary.

Much of the existing work on healthcare communication, risk, ethics, and environment, for example, is focused on crisis communication and thus deals only with urgent threats to life. A recent issue of *Technical Communication Quarterly* (Frost et al. 2021) begins to push at our understandings of what constitutes urgency in its examination of unruliness and what sorts of urgent problems unruliness can address in the context of healthcare and technical rhetorics. In that special issue, Kimberly C. Harper (2021) shows how ethos itself is a term that is constructed differently depending on positionality and identity, and Jamal-Jared Alexander and Avery Edenfield (2021) offer cases showing how marginalized peoples—for whom urgency may be omnipresent—navigate medical institutions. They likewise supplement existing evidence that those who are disenfranchised by biomedicine are those most likely to turn to alternative therapies (Derkatch 2016; Frost and Eble 2020). Peter Cannon and Katie Walkup (2021) address the reification of healthcare inequities in mental health, an entire arm of healthcare often falsely cast as less than urgent. McKinley Green (2021) interrogates how software can close off space where users otherwise might have been able to talk more openly about health status in his analysis of PrEP and HIV disclosure narratives on Grindr, thus creating new urgencies. Gina Kruschek (2019) adds a focus on stigma and the ways cultural urgencies surrounding (a lack of) disclosure develop. Hua Wang (2021) shows how economic urgency provokes subversive practices in her study of how Chinese mothers literally capitalize on their motherhood status to make ends meet. Cana Uluak Itchuaqiyaq and Breeanne Matheson (2021) encourage technical communication to move beyond decolonial metaphors, arguing that those who use the term *decolonial* should be doing active, or urgent, decolonial work—not metaphorical decolonial work.

In the same way these recent conversations reframe how we understand urgency in technical communication, my analysis of the Deepwater Horizon Disaster likewise goes beyond what is traditionally understood as crisis communication (and beyond what is traditionally understood as health communication) to include longer-term effects. I revise what constitutes *crisis* by rethinking what constitutes urgency or exigency as a corollary of *efficiency*. Our understandings of efficiency are based on an assumed time frame, and our understandings of urgency are bounded by this paradigm. If we consider different temporal configurations, our understandings of urgency and crisis are consequently affected.

Further, this particular study is unusual in that it considers environmental disasters as a catalyst for health risks and questions the role of efficiency in identifying those health risks and producing healthcare communication about them. Health risks, like environmental crises themselves, may be constituted differently by revised understandings of efficiency and its supporting terms. In some places, legal rhetorics set out temporal limits on our understandings of health effects very directly, as in the Deepwater Horizon Disaster Settlement's instruction that the time limit for bringing suit after a related diagnosis is four years. In other places, the allowable time frame of cause and effect when it comes to environmental health concerns is assumed.

On a larger scale, this book critiques contemporary instances of sex- and gender-based injustice in technical rhetorics (Frost and Eble 2015) with the goal of moving toward social transformation. As a technical communication scholar, I have found that many people outside the field (and some within) consider technical communication to be "neutral" or "objective." As a result, I became interested in questioning what precisely causes a person to believe a piece of communication is objective. Further, I became invested in disrupting these notions of objectivity. Technical documents are as situated as any other communication, but this common perception of objectivity—the notion that they *don't* persuade—actually means they have particular power *to* persuade. They also can be especially difficult sites for cultural critique. I have found, as well, that disaster communication tends to sponsor an urgency to swiftly and uncritically accept rhetorics of efficiency; as such, it is especially important to apply critical approaches—in this case, apparent feminism—in disaster scenarios.

I consider this book to be no exception to the statements I've made above. While I will do my best to be as descriptive (rather than evaluative) as possible in some passages, perhaps particularly in those places where I draw on my experiences as an investigative journalist, I also never abandon the idea that the information I present here is situated. In the words of Owens (2015, ix), "As a feminist rhetorician, I take seriously the obligation to provide context for readers about my own position and experience." Thus, this book includes what may seem at initial observation to be two distinct writing styles—one more narrative, based in experience, and one more scholarly, based in "traditional" research. My experiences affect the way I view the world, and so I work to make them apparent rather than hiding my writerly voice behind a veil of false objectivity. And yet, I also recognize that objectivity functions as a style and not always as an absolute, and so I am unwilling to completely leave its trappings behind. Finally, I also make a habit, throughout this book, to present what I think of as generative critique; that is, I attempt to offer a variety of new directions rather than suggesting that we replace one paradigm wholesale with another.

One example of this is an intervention into the ways risk communication scholarship generally looks. Studies of risk communication often attempt to find communication breakdowns prior to disaster events to prevent similar future disasters. For example, noted risk communication expert Beverly A. Sauer (2010)—who has done work on the Deepwater Horizon Disaster—often examines what is missing from risk communication documents; she finds the gaps that could prove useful in developing better risk communication in the future (Sauer 2003). Paul M. Dombrowski's (1994) work concerning the Challenger disaster is another well-known example of risk communication critique aimed at improving future practice. Most risk communication scholarship, especially that within technical communication, is explicitly aimed at working to prevent disaster in the future. In other words, it examines risk constructions with an eye to mitigate future risk—an unsurprising turn for scholars aiming to ensure the relevance of their work and also an admirable goal that is undoubtedly in the public interest. While I am happy to claim this as a tangential goal of the present text, I am working with risk not as a central term aimed at preventing future disaster but rather as a way to think in more depth about the social and cultural constructions of risk toward strategies for helping us conceive risk, disaster, and crisis differently. My concern is less with the business cultures that produce risk communication or official preparedness practices than it is with tracking how local, regional, national, and international cultural groups use communication to shape understandings of risk and disaster after the disaster has already been recognized as such. This work, too, can function in service of the public good and social justice. Altering our understandings of the common events in our lives is a necessary prerequisite to the sorts of specific, material change urged by many risk communication scholars.

FEMINIST TECHNICAL COMMUNICATION

Our point of departure for doing the work described above must be a shared understanding of the field of technical communication. Technical communication has existed as a formalized discipline for several decades,[1] but its explicit engagement with feminisms and related social justice and cultural studies approaches has been more recent. Before proceeding, it is imperative to understand the foundations from which this book's theoretical approaches emerge; this has the added benefit of also mapping the conversations this scholarship—including my apparent feminist methodology—speaks and contributes to. In what follows, I offer a particularly situated history of technical communication as a field, followed by a history of feminist interventions into technical communication.

In chapter 2, I explain apparent feminism. But here, I provide an organic enactment of this approach even before explicating it. Offering a history of a field is not a neutral act, and my approach to doing so is not traditional. We must pay attention to the development of disciplinary histories and must also strive to add origin stories that are inclusive of diverse perspectives—particularly the perspectives of women, because they have been so long excluded from traditional origin stories. Here, I do my best to avoid essentialist perspectives while at the same time establishing a flexible, permeable, temporary foundation from which to work. I strive to work across alleged disciplinary, national, and cultural divisions. I contend that all this work can be attended to by making feminist identities apparent whenever possible, by hailing allies in social justice work and recognizing their valid reasons for not self-identifying as feminist, and by constantly reimagining the purposes and functions of efficiency as a disciplinarily valued term. Most important of all, though, I recognize this as a partial history and as a beginning. This sort of apparency is valuable because it makes explicit the underlying inequities and inefficiencies that we must now tend to.

Thus, this history introduces problems. This is a deviation from what histories are supposed to do, which is to create a single, contained, cohesive narrative so that all readers operate from a shared understanding. It feels much nicer to tie a metaphorical bow out of any loose ends than to linger in uncertainty. However, I choose to introduce problems—or, rather, to make existing problems apparent—because I recognize that the methodological favors I will ask farther on are not easy, and a certain amount of discomfort may be a useful preparation. This approach doesn't solve every problem, but it does provide a method of engaging in and making apparent important conversations within technical communication that might not always be made explicit in other places. I seek to animate N. Katherine Hayles's (1999, 12) "Platonic forehand," which moves us away from simple abstractions and "evolves a multiplicity sufficiently complex that it can be seen as a world of its own." This is one way to say that this book does not seek to make the relationships it describes less complicated but rather to make apparent the complexity and importance of the relationships among technical communication, feminisms, rhetorics, efficiencies, and social justice as well as health and medical communication, technologies, and environmental rhetorics.

(Some of) Technical Communication's Origin Stories
In this section, I introduce several possible re-tellings of technical communication's origin stories—some of which coincide with familiar origin stories for the field of rhetoric. In doing this work, which follows work like that of Carolyn Rude (1979), Angela M. Haas (2007), and Jay Timothy Dolmage (2014), I seek to produce a more efficient understanding of technical communication's disciplinary history. This means that I am setting out to hail more diverse audiences—perhaps especially women and people of color—with the origin stories I will make apparent. In other words, these narratives may seem to the reader to be problems, as it is impossible in some cases to reconcile the stories I tell with the discipline's traditional stories. My goal is to make these stories explicit rather than provide all the answers about what to do with them.

I suggested above that technical communication is a young field. I based this suggestion on my observation that scholars before me have marked the beginning of the field of technical communication (though not the beginning of the practice of technical communication) based on the existence of the familiar term *technical writing*. Teresa Kynell (1999) reports that textbooks that use the term *technical writing* in their titles first began to appear in the mid-1920s. This narrative coincides with the rise to power of engineering departments in universities (McDowell 2003).

However, rooting the discipline in this singular origin story is a problem; it is inefficient in that it obfuscates more diverse understandings of what technical communicators can be and can do. I suggest now that we might imagine technical communication as much older than the particular history associated with engineering implies. Further, if we believe technical communication is much older than this, we might also believe it is a field with transdisciplinary roots, though those roots have often been rendered unapparent. Even in the most common origin stories—those tied to engineering—it is fairly apparent upon close inspection that the 1920s did not give rise to something new but rather repositioned an age-old practice in a new, professionalized way. The practice of technical communication—though it has not always been so named—is an ancient tradition that focuses on reaching a specific audience who is in possession of a specific body of knowledge. Technical communication and rhetoric—as disciplines and practices—are intricately connected; neither encompasses the other.

These imaginings are one possible remediated version of technical communication's history. However, a focus on any one disciplinary history for technical communication necessarily leaves out others. I recognize that this critique applies to my own work. I will not have the time, space, or knowledge to tell all—or even very many—of the possible origin stories that have affected the trajectory of the discipline of technical communication. This historiographic project must be a continuing endeavor. Nevertheless, this section provides a space to start such a project of reimagining. I begin this section, then, by asking: what origin stories are left out when we focus narrowly on the existence of the term *technical writing* as a marker of the discipline's existence? In answering this question, I have focused on some of the possible origin stories that are most important to make apparent from a feminist perspective.

One possibility for relocating our origin story might be to follow the work of scholars like Susan Rauch (2012), who argues that technical communication's beginnings should be tied to the work of female writers in medieval times. Rauch (2012) suggests that Hildegard von Bingen, a writer of medical and scientific texts, has not been revived and reintroduced as a technical writer in the same way as other medieval figures, one example being Geoffrey Chaucer. However, when we do consider von Bingen as part of technical communication's origins, we find that this inclusion alters our understanding of the discipline itself. For example, Rauch suggests that understanding von Bingen as a part of technical communication's history opens up opportunities today for us to consider practical approaches to women's influences on health and safety research and to point out connections between healthcare writing and related fields (397). By including von Bingen as part of technical communication's origin story, we stand to gain new understandings of the field's interdisciplinary nature, responsibilities, specialized skills, and important conversations. Of course, we also introduce many problems, including leaving out technical communicators who preceded von Bingen chronologically. I purposefully resist ordering the stories I tell here according to chronology because I do not want to imply that the stories I tell are comprehensive. I am more concerned with setting priorities for apparency than in reinforcing linear notions of time and history.

We could figure ancient Greece as part of technical communication's origin story—a move perhaps more common to rhetoric, but then, I mean to trouble the boundaries that supposedly separate technical communication from rhetoric. Specifically, Greek scholars Sappho and Aspasia have inspired historiographical work in the field of rhetoric and composition (Bernard 1999; Glenn 1994; Jarratt 1998, 2002). By counting Sappho and Aspasia as technical communicators as well as rhetoricians, we stand to widen the discipline in ways that will allow practitioners and scholars to listen to more voices from across time as well as space. Further, claiming Aspasia and Sappho as technical communicators introduces a narrative in which technical communication exists and develops hand in hand with rhetoric.[2] Although many modern scholars already span these fields and contest their separation, this is an origin story that might create more space for today's transdisciplinary scholars to do dynamic and important work that does not always fit neatly into currently existing disciplinary categories. However, again, this revised origin story introduces many problems. By beginning with Sappho and Aspasia, we leave out rhetors whose work was not recorded or reported. And, of course, we introduce a length of time so extended that we can never hope to come close to producing a representative chronicle of our disciplinary history.

Reimagining origins, though, is not just about people; it's also about sites and disciplines. We can work around the constraints introduced by technical communication's engineering origin story by revising our beliefs about our intellectual foundations. Francesca Bray's (1997) gynotechnic inquiry, applied to digital rhetorics and technology studies, demonstrates that other traditions are not only possible but have been thriving. For example, Haas (2007) argues that the term *digital rhetorics* refers first to rhetorical artifacts produced by digits, or fingers.[3] Thus, digital artifacts and technologies are—and long have been—parts of various technical systems that produce ideas about gender relationships. By embracing wider understandings of technologies and digital rhetorics, apparent feminists can draw upon and dialogue with a variety of scholar-practitioners not traditionally recognized as technical communicators, including Indigenous rhetorics scholars, cultural anthropologists, architects, linguists, archaeologists, literature scholars, artists, cultural studies scholars, and many more. This particular apparent feminist origin story is far more inclusive than the traditional engineering-oriented origin story and as such hails much larger and more diverse audiences. It is much more efficient in imagining the possibilities offered by this diverse field.

(Some of) Technical Communication's Organizing Concepts

What if we imagine technical communication as a socially just endeavor that is always necessarily a tool of the oppressed? This reimagining of technical communication means that we include more voices, stories, goals, and epistemologies in our understandings of what technical communication can be and do as well as what it *should* be and do. For example, a version of technical communication that focuses on rhetorics of the oppressed might include the story of Nujood Ali, a young girl who engaged in persuasive technical communication to convince a Yemeni judge to grant her a divorce from her abusive husband (Ali and Minoui 2010). Ali has since written a book—a technical composition—in an effort to increase the efficiency with which her story is disseminated. This new version of technical communication also might include the story of Nakato Juliette, a Ugandan mother who created a video to promote a cooperative wherein she and other women became jewelry makers to earn a living rather than prostituting themselves (Juliette 2011).[4] We might recognize these women—who do not publicly profess to be feminists but who are certainly engaged in social justice endeavors that benefit women—as technical communicators by revising our disciplinary efficiencies and obligations. This recognition permits us to contribute to their causes and to learn from their tactical interventions into systems of oppression. It helps us remember that technical communication does not only happen in academia.

However, recognizing technical communication as a tool of the oppressed also requires technical communicators to interrogate the effects of incorporating these narratives. For example, how do we identify a party or a person as oppressed, and who makes decisions on what constitutes oppression? What risks might women like Juliette and Ali face because of the apparency associated with inclusion? What specific contexts, histories, and local practices might be obscured by the presence of these women rather than others as part of technical communication's histories?

An excellent approach to thinking about technical communication of the oppressed and technical communication beyond academia is Cecilia Shelton's work on marginality (2019a). Shelton argues that her theoretical approach, "A Techné of Marginality[,] positions technical and professional communication theorists and practitioners to recognize the ways in which Black communities, and particularly Black women, have always, already done the unpaid labor that builds the [technical] communication infrastructures for equity, inclusion, and freedom" (abstract). She expands on this approach in her pedagogical article "Shifting Out of Neutral" (Shelton 2019b) by grounding her inquiry in Black feminist theory and positioning it as always already pivotal for technical communication and rhetoric:

> One of the central and critical questions that Black Feminist theory poses to those of us who want to do social justice work in technical and professional communication is, "How do we decenter whiteness (and other privileged identities) to insist on a more intersectional analysis of oppressive

systems and the activism that disrupts those systems?" This article takes up that question by arguing for the inclusion of two themes which have been, so far, largely overlooked in technical and professional communication scholarship: the invisible labor of being Black women in the field of technical and professional communication and the significance of our bodies as texts in our classrooms. (18–19)

Shelton's work responds to some of the questions I raise above by pointing out that self-identification is an important method for determining whose marginality might point to associated expertise and that those on the margins are always already running some of the risks white feminists worry about exposing. Apparency, in this case, can be co-opted to help recognize infrastructure that has long existed without being recognized.

Indeed, including and centering Black experience is an astoundingly obvious and yet still too often glossed-over response to recognizing technical communication as a tool of people who have suffered marginalization and oppression. In 2020, the Conference on College Composition and Communication (CCCC) published a position statement from the Black Technical and Professional Writing Task Force (Mckoy et al. 2020).[5] The task force wrote that "Black technical and professional communication is defined as including practices centered on Black community and culture and on rhetorical practices inherent in Black lived experience" and that it includes academics and practitioners (para. 2). It advocated including and amplifying Black practices and the work of Black scholars, and the task force immediately put words into action by providing a detailed thematic list of suggested readings. Later, the same group of scholars who comprised the task force—Temptaous Mckoy, Cecilia Shelton, Donnie Johnson Sackey, Natasha N. Jones, Constance Haywood, Ja'La Wourman, and Kimberly C. Harper—offered the field a further tour de force in definitional work and inclusion in a complete special issue of *Technical Communication Quarterly* (Mckoy, Shelton, Sackey et al. 2022). Black technical and professional communication (BTPC), the authors wrote in their introduction, "is not a niche or add-on subfield of the discipline of TPC [technical and professional communication], even though it has traditionally been treated as such. BTPC is an important and integral part of TPC and foundational to understanding how TPC is taken up, applied, theorized, and shaped in culturally sustaining and contextual ways" (221).

This special issue demonstrates both that Black rhetorical practices have always already been central to the work of technical communication and that the inclusion and centering of marginalized knowledges is welcome and inevitable. The issue's conclusion forwards the same main argument as does this book: that intersectional and social justice approaches to TPC centering the voices of those who have been historically marginalized moves us toward better communicative practices.

Challenges and Professionalization

A potential problem for apparent feminist reimaginings of technical communication's origin stories is the possibility that some scholars might suggest I am promoting an agenda in which technical communicators would lose the social capital associated with special expertise. The question of how (and if) to professionalize—that is, to engage in strategies that persuade others of the specialized expertise and professional identity of this discipline—is an ongoing debate in technical communication circles (Carliner 2012; Coppola 2012; Davis 2001; Faber and Johnson-Eilola 2002; Johnson-Eilola 1996; Kynell-Hunt and Savage 2003; Savage 1996, 1999, 2003, 2010). I understand the professionalization debate as one manifestation of the ongoing argument about the relationship between diversity and efficiency. The notion that increased diversity of origin stories necessarily means decreased social capital for technical communicators is erroneous, as Shelton shows. Rather, I follow Shelton and Savage in believing that many contemporary understandings of professionalization are based on "a modernist agenda which is no longer appropriate for a field of work for which modernist notions and practices are less and less relevant or useful" (Kynell-Hunt and Savage 2003, 170).

All of the origin stories I have mentioned above—and many more that I haven't—are narratives that can help expand researchers' and students' understandings of the discipline of technical communication. By thinking differently about where technical communication comes from, what it is constituted by, and who is creating it, we can think differently about the places where it might go and the tasks it might confront. By imagining a more diverse group of actors, we can imagine more diverse audiences, scholars and practitioners, and students. This is a step toward helping all these actors to "reconceive the profession as one that can be practiced in alternative ways" that privilege integrity and social justice (Savage 1996, 310). Just by drawing inspiration from the very few possible origin stories I have mentioned, we open up new realms of immense transdisciplinary possibilities in terms of future research and perspective. We also introduce a vast repertoire of new problems and challenges, not the least of which is resistance to disciplinary change and practices of inclusion.

Disciplinary friction is, in fact, an important part of technical communication history.[6] That friction is also something that is not often written about in formalized settings—for a number of reasons, but at least in part because so many technical communication scholars hail from multidisciplinary English departments. Those departments are the very places this friction lives, and the same departments are made up of the people we need to get along with every day to do our work, best mentor our students, and achieve tenure and promotion. In the introduction to a reprint of Robert Connors's (1982) history of technical writing, R. Gerald Nelms (2004, 4) writes explicitly of how the history records that "a major obstacle to progress in technical writing instruction—one that handicaps all writing instruction—is the dominance and prejudices of English department literature faculty." Nelms points out that Connors wrote his history without having tenure and describes him as courageous in intervening in elitism that interfered with instruction in English departments. Connors himself wrote of Samuel Chandler Earle, who believed "the problem of a cultural split between English and engineering teachers" was significant.[7] "He condemned the attitude of English teachers who saw engineers as philistines, to be proselytized to about the superior virtues of culture and literature over engineering" (2004, 6).

From the privilege of tenure, I can confirm that disciplinary tensions between literature scholars and technical communication scholars do continue to impact curricula—perhaps particularly when it comes to decisions about who we should hire to best serve students, which then impacts our course offerings—and, further, that different constellations of disciplinary understandings do so as well.[8] For example, I earned my PhD at Illinois State University and left that institution with an understanding of rhetoric and composition as a widely recognized discipline and technical writing as a subsidiary of rhetoric, though all three are related and different understandings of technical communication and rhetoric might reverse which of them is the umbrella term; put another way, I generally believed most technical communication scholars also considered themselves rhetoricians, whereas the reverse was not necessarily true. I took my first tenure-track job at East Carolina University, where I was hired as a technical and professional communication specialist and expected to caucus in a group of similarly situated people—some of whom did not consider themselves rhetoricians—which did not include a separate group of rhetoric and composition specialists who caucused on their own. While nearly everyone in these two specialties seems to recognize that the divisions are somewhat arbitrary, the divisions remain because they are also political.

The gendered language and ideas that emerge throughout Connors's history (that is, both the language he reports and language he uses/reflects) are worthy of note. Dissatisfaction with the discipline is "shrill" (2004, 10), and English teachers are accused of being too effeminate. It is impossible to miss the association of literary, humanistic work with the feminine and of technical, vocational work with the masculine. This association lives on today, with an implicit understanding that masculine work is more highly valued. This understanding goes beyond cultural value; work that is conceptualized as masculine is better paid. As many technical communicators—also often trained in rhetoric and composition—know, the history of composition instruction in the United States is rife with gendered assumptions. From the lament that "the first-year composition courses were 'given to just about anybody who would take [them] . . . faculty wives, and various fringe people, are now the instructional staff' " (quoted in Crowley 1998, 119) to a variety of studies of the feminization of composition (Holbrook 1991; Schell 1998), we know where this history comes from and where it leads. It took years before composition gained some recognition as its own, professional discipline, and it still lacks prestige—at least as measured in dollars—compared with many of the disciplines it serves. Technical communicators have worked hard to avoid a similar fate, even as many technical communicators—perhaps especially the feminist ones—feel a deep kinship with composition studies. The result is a somewhat fractious approach to professionalism and its trappings.

Feminisms in Technical Communication

Professionalization is a common topic among technical writers, and informal markers of licensure are useful historical points for telling histories. Some would say "technical writing finally became a genuine profession as wartime technologies were translated into peacetime uses" and "the demand for [technical writing] courses rose dramatically as the colleges were deluged with returning veterans after 1945" (Connors 1982, 341). If this history is to be believed, then technical communication was growing up as a field just before the time when second-wave feminism was gaining power—the first wave, which focused largely on property rights and suffrage, having ebbed by the early 1920s. The second wave, often said to have begun with the publication of *The Feminine Mystique* (Frieden 1963) and certainly associated with the Civil Rights movement, shifted attention to identity and gender roles. Women began to question the notion that being a wife and mother was the only path to success as a woman. The second wave gave rise to various kinds of feminisms that were sometimes in conflict with one another;[9] for example, cultural feminists' belief in valuing traditionally female roles could sometimes clash with liberal feminists' injunctions to respond to stereotyping with resistance. It is the second wave when feminist began to wrestle in earnest with the concept of subjectivity, which, "with its explicit universality but implicit masculinity, creates a dilemma for feminism" (Meagher and DiQuinzio 2005, 3). Patrice DiQuinzio and Sharon M. Meagher (Meagher and DiQuinzio 2005) explain that feminisms are trapped into arguing for equality of the sexes by denying sexual difference but then must rely on sexual difference to analyze the unique experiences of women. Second-wave feminisms and all their complications and perceived excesses, I have discovered, are often the feminisms early-career college students are still responding to; they are the feminisms that are characterized—or, more accurately, caricatured—in popular media. And it is at what is typically considered the end of the second wave that explicitly feminist interventions into formal technical communication literature began.

Mary M. Lay's (1989) article "Interpersonal Conflict in Collaborative Writing: What We Can Learn from Gender Studies" is widely regarded as the first explicit engagement of technical communication with gender studies.[10] In this piece, she transfers gender studies' knowledge of the ways gender perceptions affect relationships to the domain of technical writing and offers strategies for helping technical communication students understand the limitations of gender roles and better collaborate. As is often the case with both cutting-edge and teaching research, Lay's work was not immediately taken up. However, the same journal, the *Journal of Business and Technical Communication* (*JBTC*), evidenced its commitment to understanding technical communication through a cultural lens with a special issue two years later.[11] In fact, as noted by Isabelle Thompson (1999, 155) in her qualitative content analysis of journal articles from 1989 to 1997, "most journal articles about women and feminism in technical communication appeared in special issues devoted to those topics."

The 1991 special issue of *JBTC* promotes a cultural turn in technical communication. This cultural turn, which is an important prerequisite for the reimagining of efficiency that I suggest, was not widely taken up, as suggested by the much later return to the idea by J. Blake Scott, Bernadette Longo, and Katherine Wills (2006).[12] This special issue provides important foundations for work in feminisms and cultural studies; it also points to what some might consider a conflation of these two theoretical approaches. For example, Lay (1991) suggests a redefinition of technical communication that considers cultural issues, most notably issues of gender. She relies on technical communicators' understandings of social constructionism to combat and make visible scientific positivism in technical communication artifacts. Diane D. Brunner (1991, 409) encourages recognition that "we and our students operate within a culture in which domination/subordination is produced and reproduced" and that, embodied as we are, this creates ideologies in which some people are affirmed and others are cast out. Others in the issue advocate revision to static conceptions of female cultures and resistance to auto-colonization (Carrell 1991; Flynn et al. 1991).

Notable in this same special issue is an article by Elizabeth A. Flynn, Gerald Savage, Marsha Penti, Carol Brown, and Sarah Watke (1991). This article stands apart in its attention to and explicit naming of gender studies. Flynn and her coauthors specifically advocate bringing together composition studies, gender studies, and technical communication as a methodological approach. The authors find unrecognized misogyny in their studies of engineering students. They frame feminist gains in the field as "fragile and provisional," suggesting that "there is little evidence that women are aware of the potentially threatening climate in which they operate daily" (460). These authors, then, come together in this special issue of *JBTC* to suggest that feminist approaches to technical communication are a necessary remedy to the field's unrecognized male domination.

The journal *IEEE Transactions on Professional Communication* furthered this project with a 1992 issue devoted to the effects of gendered assumptions on understandings of rationality. Elizabeth Tebeaux and Mary M. Lay (1992) engage in a historiographical recovery of English Renaissance–era technical writing for women; Kathryn Neeley (1992) explicates a history of women mediators in the eighteenth and nineteenth centuries, and Beverly A. Sauer (1992) argues that gendered assumptions about male ways of thinking affect mine safety management. L. J. Rifkind and L. F. Harper (1992) assert a paradox between sexual harassment policies and the necessity of interpersonal relationships in the workplace, and Sherry A. Dell (1992) draws in communication theory in a rhetorical analysis of the "glass ceiling." Stephen A. Bernhardt (1992) and Deborah S. Bosley (1992) separately engage issues of gender in visual design. Notably, several of the authors in this special issue are among a small group of scholars whose work consistently shows up in the disciplinary special-issue space that seems to be reserved for feminist issues.

Two years later, *Technical Communication Quarterly* (*TCQ*) expanded on feminist approaches to technical communication with an issue that "explores gender as a social force that shapes and is shaped by professional communication practices and readerships" (LaDuc and Goldrick-Jones 1994, 246). Linda LaDuc and Amanda Goldrick-Jones (1994) invoke the power of feminism's ability to take on multiple theoretical and political positions. This multidimensional approach reflects an understanding of the importance of "forsaking the comfort of even a single feminist method or 'truth stance'" (249) in favor of embracing diverse methodologies that avoid feeling "mechanical" (Christian 1987, 53) and instead make otherness more apparent. Laura J. Gurak and Nancy L. Bayer (1994) and Sauer (1994) engaged in this kind of complex work by writing articles that describe a variety of feminist methodological approaches (and resulting implications) to their subjects rather than limiting their investigations to a single methodological approach.

Some of the work in this special issue—especially the articles by Jo Allen (1994), Bosley (1994), and Susan Mallon Ross (1994)—also continues the aforementioned conversation about the field's unmarked maleness. These articles suggest that resistance to hegemonic, masculine notions of technical communication has already begun and that such resistance gains power from interdisciplinary awareness. Both Allen and Bosley point to ways of challenging and marking this invisible valuing of maleness; Allen finds that both women and men are already subverting traditionally gendered modes of constructing authority in technical documents, while Bosley showcases attention to the rich value of perspectives from other disciplines. Most notably, she cites gender studies scholar Susan Bordo to situate the "masculinization of thought" (Bosley 1994, 297) in technical communication. Like Bosley, Ross looks to sources outside the discipline for additional insight; she pushes for intercultural studies such as her own on the interactions between a Mohawk community and the Environmental Protection Agency. In so doing, she provides an example of how feminist concern with other injustices—namely, racism and environmental oppressions—can inform broader understandings of the applicability of feminism to a field like technical communication. Feminism and social justice agendas, in other words, are often symbiotic.

Social justice and social change have been advocated by many technical communicators; the 1997 *TCQ* special issue recovers histories of women technical communicators and questions the absence of such histories. Katherine T. Durack (1997) begins by suggesting that women's work in technical communication has been overlooked because the field has been understood as the domain of men and because historians have tended to internalize that belief. Elizabeth A. Flynn (1997) and John F. Flynn (1997), among others, begin to remedy this situation by paying attention to the mapping of feminisms in technical communication and by engaging in the recovery of domestic sciences and technologies—like grocery shopping, cooking, and bread making—as technical communication practices.

Technical communication journals have largely left feminisms and gender studies behind as a named topic for special issues since 1997, although complementary and related approaches have sometimes been evident. For example, *TCQ* published a special issue on "New Directions in Intercultural Professional Communication" in 2014 that included articles with cultural studies approaches, as well as a special issue on "Tactical Technical Communication" in 2017. This change in the focus of special issues mirrors a larger shift in the field; "technical communication scholars' interest in feminism and women's issues has declined over the past 15 years" (Thompson and Smith 2006, 196).

While the special journal issues I reviewed above may provide the most systematic, discipline-sponsored engagements with feminisms by technical communicators, these issues are not the only examples of feminist technical communication work. For example, feminist technical communicators whose work has appeared in more solitary fashion have taken up the historiographic project of the 1997 *TCQ* special issue. However, it is important to understand that the historical injustices that are to some small degree remedied by this work also provide windows into contemporary exigencies: who is being left out in the histories and recoveries we are currently writing? Who is being left out of this very review of literature? Kyle P. Vealey and Alex Layne (2018) argue that scholars cannot treat ontology as just an abstract topic of inquiry but rather must consider the ways our scholarly practices create realities. Using the example of scholars of object-oriented ontologies and their tendency to elide and ignore their feminist influences, Vealey and Layne argue for more responsible and ethical citation practices by putting forth a theory of reverberation: any piece of scholarship and its attendant citation practices create ripples outward that affect past, present, and future.

In an effort to sponsor ethical reverberations as I create this brief history, I now devote some space to discussing the work of technical communicators whose feminist or gender-based scholarship has appeared in more solitary, self-sponsored fashion rather than appearing in the special issues discussed above from the late 1980s to 2012. Maryanne Z.

Corbett (1990), Sherry A. Dell (1990), and Jeannette Vaughn (1989) all address sexist language in technical documentation. Ann Brady Aschauer (1999), Lee E. Brasseur (1993), and Mary M. Lay (1993) interrogated the intersections of gender and technologies. Evelyn P. Boyer and Theora G. Webb (1992) and Maria de Armas Ladd and Marion Tangum (1992) looked to diversity and difference as guiding principles in feminist thought in technical communication. A number of other scholars in the early 1990s also did important work on the subjectivities of technical communication and on the importance of feminist methods and of having female perspectives in the profession (Brown 1993; Coletta 1992; Dragga 1993; Sauer 1993; Tebeaux 1993).

As part of this history of individual scholars, my assertion of a lack of feminist work in technical communication in the fifteen+ since 1997 (Frost 2016) does not constitute a complete absence. Rather, it signals a waning of interest at a kairotic moment that is, from my perspective, politically unfortunate. However, this waning of interest—and my mentioning it—is not meant to discredit the important work of those relatively few scholars who have continued to publish at the intersection of feminisms and technical communication. For example, Gail Lippincott (2003) examined Ellen Swallow Richards's rhetorical development of an ethos that allowed her to do work with her experimental food laboratory, the New England Kitchen. Brasseur's (2005) historiographic work on Florence Nightingale's persuasive use of rose diagrams to advocate for government reform of sanitary conditions in hospitals points out that Nightingale was a talented administrator, statistician, and technical communicator. Some technical communication scholars have also taken up cultural studies approaches with explicit feminist components. Jeffrey T. Grabill's (2007) work focuses on the ways information technologies penetrate and shape everyday lives, and he encourages emancipatory action on the part of citizens—especially women and people from economically disadvantaged communities. Meanwhile, Angela M. Haas, Christine Tulley, and Kristine Blair (2002, 247) complicate constructions of women's and girls' relationships with technology and technical communication, arguing that it is dangerous to "presume that 'going online' somehow alleviates gender inequity and power imbalance," especially given the traditional masculine gendering of technology. In response, they propose feminist methodological alternatives to male-centered models for "mastering" technical communication and technology. In sum, these scholars—and others I do not have the space to discuss at length (e.g., Koerber 2002; Lay, Monk, and Rosenfelt 2001)—have kept many threads of feminist inquiry alive and thriving in technical communication.

Since 2012, interest in the social justice movement in technical communication and, thus, intersectional feminist work as well has gained some steam. However, feminist work in technical communication is still regarded by some as a box that has been ticked—a sentiment I heard expressed directly by a well-respected technical communication scholar in the audience at the plenary session of the 2012 meeting of the Council for Programs in Technical and Scientific Communication. As Kate White, Suzanne Kesler Rumsey, and Stevens Amidon (2016, 29, original emphasis) put it in their update to Isabelle Thompson (1999) and Thompson and Smith's (2006) aforementioned work:

> In our initial analysis of textbooks and journals in the field, we were stunned to discover that an implicit message seems to be inherent in the published discourse of our field that issues of gender and feminism in the workplace or in our business and technical writing classrooms are a minor concern. In many ways, this published discourse seems to be doing little to challenge the insidious notion that the workplace is neutral and nongendered. This implicit message makes little sense to us, given the presence of dozens of scholars and teachers in our field we meet at conferences who are obviously interested in these issues. For instance, the Association of Teachers of Technical Writing (ATTW) Conference has created an event for women to discuss issues that affect them professionally including obstacles to success and both feminist research and administrative practices. However, we couldn't ignore what the written record seemed to be telling us. While examining our own teaching practices, our textbooks, and the leading journals in business and technical writing studies, we are disturbed to find that very little seems to have changed in the past 25 years. We were surprised and dismayed not at what we found in textbooks and professional journals, but [at] what we *didn't* find.

Put another way, White, Rumsey, and Amidon observed the same pattern I did in 2016: an apparent waning of interest in feminist work in technical communication literature. At the same time, they point out that this waning interest is not reflected in the field's less formalized work. Regardless, in the absence of widely apparent special-issue–sponsored approaches to feminist technical communication, the burden of gender-based and cultural work is born by a limited number of dedicated scholars who must work for apparency. In most cases, these scholars tend to be made responsible for doing this apparency work in addition to all the work expected otherwise.

These individualized approaches continue within a network of feminist technical communication scholars who more broadly embrace social justice as a vital foundation for future work in the field—and who increasingly combat the isolation of individual scholarship on social justice issues by coauthoring. Much of this work has been made possible by

Haas's (2012) argument for intersectional approaches to race, rhetoric, and technology. Since then, a number of other works have been published that both directly and indirectly engage with feminist and gender studies approaches to technical communication as part of a larger cultural studies–savvy body of work.

Some recent scholarship directly points to concerns for women as producers and consumers of technical communication. Kathryn R. Raign (2018) recovers the history of Enheduanna, the first woman writer, and uses that recovery to cast doubt on histories that suggest that men developed and honed persuasive and technical writing. Patricia Sullivan and Kristen Moore (2013) track infrastructural mentoring practices and needs for women in engineering, and Jennifer C. Mallette (2017) writes about recruiting and retaining women in engineering fields. Emily January Petersen (2014) discusses the formation of motherly identity through blogging. Valentina Rohrer-Vanzo, Tobias Stern, Elisabeth Ponocny-Seliger, and Peter Schwarzbauer (2016) examine the effect of gender on assembly documentation, with a specific focus on male technical writers producing documentation for female consumers. Lehua Ledbetter (2018) examines a group of women who produce tutorial-style videos and explores their uptake of and connections to (or lack thereof) feminist approaches.

Other scholars take more systemic or philosophical approaches, such as Petersen's (2019, 37) technical and professional communication workplace study that showed the gendered difficulties of navigating power structures, which are "loose, unarticulated, malleable, and negotiable." Jared S. Colton, Steve Holmes, and Josephine Walwema (2017) take up the work of feminist philosopher Adriana Cavarero to complicate the way ideas like Michel de Certeau's (1984) tactics and social justice have been taken up in technical communication through an analysis of the hacktivist group Anonymous. Natasha N. Jones (2016a) uses feminist theory and the concepts of voice and silence to offer an alternative approach to technology production and design.

Importantly, a number of scholars who engage (directly or indirectly) with feminisms have also emphasized intersectionality and feminisms' place as just one of a number of cultural approaches aimed at social justice (Baniya et al. 2019; Edwards 2018; Garrison-Joyner and Caravella 2020; Lockett 2019; Williams and Pimentel 2014). In an already well-cited 2018 article, Petersen and Rebecca Walton put feminisms into conversation with social justice work. Jones (2016b) argues for a social justice approach to technical and professional communication and articulates feminisms as part of such an approach. Jones, Kristen R.

Moore, and Walton (2016) draw on feminisms and gender studies, work in race and ethnicity, and intercultural and international approaches to professional communication to disrupt the dominant narrative/history of technical communication as a field; in so doing, they suggest that new (understandings of) pasts can create new futures. Petersen and Ryan M. Moeller (2016) take a similar approach in their treatment of antenarrative as a methodology for feminist historiography of IBM. In the sub-field of rhetorics of health and medicine, Lori Beth De Hertogh (2018) argues for an intersectional feminist digital research methodology for medical rhetoricians, particularly those working with vulnerable online communities.

Vulnerable communities are necessarily a focus of BTPC, which is always already complementary to feminist goals in its articulation of a desire for justice. Miriam F. Williams and Octavio Pimentel's (2016) edited collection offers a carefully curated diversity of scholarship on identity work in technical communication, focused on race and ethnicity but engaging many research methods shared with feminist work. Jones's (2017) study of rhetorical practices of Black entrepreneurs uses narrative as a methodological tool and shows the importance of an explicit practice of cultural empowerment. Laura L. Allen's (2002) rich description of various methods of collaborative leadership as shown through Black family reunion planning and management demonstrates that the boundaries between what is "professional" and what is "social" are not nearly as impermeable as objectivists might like them to be. Likewise, her attention to her role as a researcher through a critical race-grounded theory approach exemplifies a commitment to positionality complementary to feminist work: "I experienced the most anxiety with this project when I knew I would be attending my first reunion with the Marshall-Beaumont family without a Marshall-Beaumont reunion t-shirt. My unease led me to contact Shannon, my colleague and a member of the Marshall-Beaumont family, to ask her the t-shirt color so that I could blend in as much as possible" (238).

In a similar vein, Constance Haywood's (2019, para. 2) work on the difficulty of studying communities to which ones belongs demonstrates the importance of a commitment to care and reciprocity for academics: "I am in a position where I could replicate the very things [e.g., surveillance, digital aggression, white supremacy] that those ahead of me have worked so hard to avoid. This is why I find such importance in (re)thinking the hows and whys of researcher-community relationships." Attention to identity and affective response is a common theme for both feminist and Black technical rhetorics. Antonio Byrd's (2022) theorization of "Black technical joy" features descriptions of the rhetorical practices used by Black professionals as they navigate the software industry, and his work points to the importance of the recognition of affect in identity building. These stories "feature a range of emotions that point out the humanity of Black people and the ultimate joy they attach to the process of learning TPC: paying the risk, failing fast, trusting the process of failure" (299). Byrd points out, too, that "racialized emotions remain undertheorized in the field as research centers white experience in the workplace" (299); I would add that the same is true of feminist theory. That is, while affect is widely recognized as important in feminist works, plenty of space remains for more explicit explorations of racialized and intersectional affective encounters in technical contexts.

Jessica Edwards and Josie Walwema's (2022) brilliant historiography of the 1881 Atlanta Washerwomen's strike puts on display the affordances of *affirmative technical communication*, or rhetorical work that engages in technical communication to work on behalf of the oppressed. Essentially, this is a form of working within the system through self-determination and collective action, even though the system was not designed for the people engaging in the technical communication in question.

Queer theory, too, has recently emerged in technical communication as an important parallel to feminist work. Matthew B. Cox (2019) draws on feminist theory to help in mapping queer professional discourse. Fernando Sánchez (2019) makes a powerful critique of textbooks' cis assumptions. Avery C. Edenfield, Steve Holmes, and Jared S. Colton (2019) use queer theory to examine the tactical technical communication genre of user-generated instruction sets, in this case instructions for the self-administration of hormone therapy for trans individuals. Indeed, queer theory provides an important complement to feminist technical communication work that I will return to shortly.

Scholars who have explicitly engaged materialisms and embodied work include Kathryn Yankura Swacha (2018), who explores the concept of embodied literacy vis-à-vis feminist theory through a classroom study of a cookbook activity; she shows what a critically informed pedagogy attentive to vulnerable populations and attuned to embodied literacy might look like. Sweta Baniya, Les Hutchinson, Ashanka Kumari, Kyle Larson, and Chris Lindgren (2019, 6) argue for attention to "what our bodies go through as we engage in our research," and Kelly Medina-López (2018) grounds her cultural rhetorical understanding of embodiment in literal building. Meanwhile, Maria Novotny and Hutchinson (2019) offer a previously unexamined genre for technical communicators to consider: fertility and period-tracking software applications.

They look at examples of this genre in detail and argue that "hundreds of other gendered health applications that could benefit from further analysis and action—not to mention technological applications in general"—exist for technical communicators to take up (357). Allegra W. Smith's (2014) poster brief on tagging and filtering systems used in a mainstream and a feminist porn sub-Reddit points to the importance of understanding users—especially often overlooked populations including women—when engaging in interface design. Liberatory cookbook narratives and the ways women are treated as consumable are also subjects of critique (Moeller and Frost 2016). In returning to the importance of bodies, Novotny (2015) studies the reVITALize Gynecology infertility initiative using apparent feminism to show how both the project and the methodology rely on stakeholder input—an important contribution I take up in more detail in chapter 2. Kimberly C. Harper's (2017, 2020, 2021, 2022) body of work details the ways the fertility industry and fertility communities, despite a history of conversations about embodiment, center white-embodied experiences. Finally, Maureen Johnson, Daisy Levy, Katie Manthey, and Novotny (2015) offer a key concept statement on *embodiment* for feminist rhetorics, one that complicates the body's usual subject/object binary position.

Several book projects in the past decade have also engaged feminisms, sex, or gender and technical communication as main themes. Amy Koerber's monographs (2013, 2018) deal with women's issues. Walton, Moore, and Jones (2019) embrace feminist perspectives as part of a larger orientation to social justice and cultural awareness, as do Godwin Y. Agboka and Natalia Matveeva (2018). Earlier, and influential for feminist technical communication scholars but beyond what is generally considered within the body of technical communication work, Donna J. Haraway's books (1989, 1991) made a significant impact.

In reviewing feminist literature in technical communication, a problem of apparency arises. To illustrate: Thompson's 1999 study limited her corpus by identifying only articles that had a keyword relating to women or feminism; she surveyed 1,073 articles and came up with 40 while acknowledging that "articles not included in this analysis have made significant contributions to research about women and feminism in technical communication" (157). Thompson also points out that key terms with the root word *gender* were always coded as feminine in the articles she analyzed. This pattern betrays the common assumption that gender is something women and non-binary people do rather than something performed by everyone—a conflation that largely holds true in the field's literature today. Further, feminist technical communication scholarship has struggled to be fully inclusive of lesbian, bisexual, queer, and trans experiences. Even when engaging in feminist work, we too often assume that the subject in question is an assigned-female-at-birth ciswoman—and I know this because I recognize the patterns in my own thinking and have to work against them to disassociate from learned assumptions and flattenings of identity. Thus, finding robust representations of intersectional feminisms through keywords becomes nearly impossibly complex.

(Un)apparent Feminisms in Technical Communication

But if feminism is not apparent or readable in keywords, does that mean it is absent from the literature? I argue that this is not the case. For example, a 2012 issue of *Technical Communication Quarterly* included several articles with feminist perspectives without explicitly marking the issue as a feminist work. Sarah Hallenbeck (2012, 305), in her study of women bicyclists, argues for the complexity of the ways "extraorganizational" technical communication reshapes technologies along gendered lines, suggesting that "normalization can be resisted, complicated, and amended even after a technology becomes commonplace." Carolyn Skinner (2012) analyzes the rhetorical strategies of Julia W. Carpenter in the late nineteenth century as she navigated the competing rhetorical requirements made of her as a physician and a woman. Hannah Bellwoar (2012) uses CHAT to research a woman's navigation of medical and health rhetorics related to reproduction. All of these articles could easily be framed as productively contributing to feminist technical communication, and this issue is just one example of how feminist work is not always apparent as such.[13]

What is essentially a labeling/metadata difficulty has impactful corollaries beyond the field of technical communication. For example, when I began teaching at Illinois State University in 2008, I promptly encountered a contingent of young women, mostly from affluent families in the Chicago suburbs, who subscribed to every feminist take on popular issues we discussed (e.g., equal work for equal pay, legal recognition of sexual harassment, reproductive rights, the viability and importance of women as political candidates) but adamantly insisted they were not feminists. I become frustrated by these conversations, in particular during a semester in which I was teaching a course on the history of rhetoric from a feminist perspective; this group of young women would easily engage with anti-feminists in the class, win their point, and then promptly disavow feminism and the very scholars who laid the foundations for the arguments they had just made.

It was this experience, along with the development of many anti-choice laws related to fetal ultrasound culminating in 2012, that led me to develop an apparent feminist methodology for technical communication and rhetoric, which I will preview briefly here—since it is part of the literature of feminist technical communication—before explaining it in detail in chapter 2. Apparent feminism addresses political trends that render misogyny unapparent, the ubiquity of uncritically negative responses to the term *feminism*, and a decline in centralized feminist work in technical communication. More specifically, it suggests that the manifestation of these trends in technical spheres requires intervention into notions of objectivity and the regimes of truth they support. Apparent feminism is a methodology that seeks to recognize and make apparent the urgent and sometimes hidden exigencies for feminist critique of contemporary politics and technical rhetorics. It encourages a response to social justice exigencies, invites participation from allies who do not explicitly identify as feminist but do work that complements feminist goals, and makes apparent the ways efficient work actually depends on the existence and input of diverse audiences.

Based on the literature reviewed above, apparent feminists might make the argument that a feminist resurgence in technical communication has begun. I suggest that apparent feminists must listen to this important work while at the same being aware that this work represents a minority population of scholars and that the disciplinary trend since around 2005 should be troubling to feminist scholars.

As shown here, feminist technical communicators have long paid attention to the need for more wide-ranging feminist approaches. In brief, then, feminist technical communicators:

- Embrace a plurality of feminisms and describe myriad feminist methodologies that can support even more feminist methods
- Persist in doing important feminist work even in the absence of discipline- and journal-sponsored forums
- Work within and across gender studies, cultural studies, and social justice agendas
- Conduct historiographical research and engage in scholarly conversations about the impact of that research
- Provide critical perspectives on technologies, sciences, terminologies, and social conventions (both within and outside the discipline itself) that hide value systems wherein misogyny is supported, valued, and reproduced
- Engage in interdisciplinary scholarship and activism
- Pay attention to the importance of intersectionality in terms of oppressions, theories, methodologies, practices, and more

- Intervene in problematic actions (including rhetorics) that exist in and between public spheres, private lives, disciplinary venues, and pedagogical spaces.

In addition to and in support of these projects, intellectual and activist transmigrations—described by Haas (2008a, 57) as points of exchange "dedicated to respectful and reciprocal dialogue"—constitute an important tradition for feminist technical communicators.[14] As such, apparent feminism requires its practitioners to draw on and contribute back to interdisciplinary feminisms. In the next chapter, then, I survey feminist theories, methodologies, practices, and rhetorics—from intersectional traditions including gender studies, rhetoric and composition, progressive social criticism, cultural rhetorics, and anthropology/sociology—that should be taken up in the disciplinary conversations of technical communication. By emphasizing transdisciplinary and transmigratory feminisms, I point to spaces for intersection and argue that feminist technical communicators should pay attention to the development of disciplinary histories, including being critical of the exclusion of diverse feminisms from technical communication. At the same time feminist technical communicators work to avoid essentialist perspectives, we also must establish a flexible, permeable, temporary foundation from which to work.

A PLAN FOR *FEMINIST TECHNICAL COMMUNICATION*

As explained above, this book's main focus is on demonstrating the importance of feminist technical communication and the network of theory it relies on and contributes to. As one method of doing so, it looks at how our understandings of efficiency bear on our actions as they relate to several common feminist concerns: ecology, economy, health, and their interconnections. Although the reviews of various literatures in this chapter have established some temporarily stable understandings of these terms before moving forward, it is important to remember that—like queer time—they are concepts that are fluid and ever-changing.[15] Indeed, the same can be said for feminist technical communication. As Cheryl Glenn (2018) argues, rhetorical feminisms are changeable, responsive, tactical. A term I would add to this list, although Glenn does not dwell on it specifically, is that rhetorical feminisms are reactive—and this is not a bad thing. Specifically, rhetorical feminisms in technical communication are currently both responding and contributing to the social justice turn.

Rhetorical feminisms as reactive is an important concept that allows feminisms to respond to the perceived objectivity of technical communication as a field. Reactivity allows feminisms to be responsive to both systemic and personal instances of sexism and misogyny as well as other intersectional concerns. As Jacqueline Rhodes (2018, 90) says, "I reject . . . any feminism that doesn't include systemic analysis as well as personal liberation—an analysis that must include discussions not just of gender but also of race, class, sexual orientation, ability, and how those things intersect." My first instantiation of apparent feminism, as previously mentioned, came about as a reaction to young college women's resistance to feminism as a term but not a concept. Apparent feminism does not aim to vitrify an approach to feminist objects of critique, smoothing over the rough edges of various feminist approaches with direct heat. Rather, it advocates a slower approach—no less angry when anger is called for but deliberate, temporary, permeable. Apparent feminism is a rhetorical feminism, able to finesse, open to change, amenable to its own obsolescence. Indeed, at its heart, apparent feminism was always meant to be situational; when I coined the term, it was partly in response to an idealist desire for a world in which "feminism" would indeed be an unreasonable perspective because equity would have been achieved and no need for a focus on "femme" would exist. That world, of course, will not be achieved in the foreseeable future, and so rhetorical feminisms—like apparent feminism—remain necessary. In combination with queer theories that offer models for temporal stretching, theorization of slow crisis expands apparent feminism's effectiveness.

This book is messy—it uses multiple methods and builds on a methodology that is permeable and flexible. This is on purpose. As Rebecca Walton, Maggie Zraly, and Jean Pierre Mugengana (2015, 46) tell us: "Process-focused pieces . . . are rare in technical communication . . . Pieces that foreground not only process but the messiness of that process are not the norm . . . and technical communication scholars wrestle with anticipating and navigating messiness in designing and conducting community-based research." The authors argue that foregrounding messiness can help us, as a field, learn to better match methods with objects of study. With attention both to allowing messiness and to the layered metaphors at work (Fleckenstein, Spinuzzi, Rickly, and Papper 2008)—metaphors that speak to economy, ecology, and health—I seek to animate these ideas in this book.

In chapter 2, I offer a detailed explanation of apparent feminism as a methodology. While apparent feminism grew out of the teaching frustrations I described above and did so at a time when anti-choice legislation was on the rise, it is not limited to addressing situations having to do with reproductive rhetorics or pedagogy. Throughout this book,

I demonstrate how apparent feminism can be applied in these contexts and more. Chapter 2 takes reproductive justice as a common example to show where apparent feminism came from and to demonstrate the simultaneous messiness and productivity of that invention process.

Chapter 3 maps slow crisis onto apparent feminist contexts as a way of showing the complementary work social justice, queer theory, and apparent feminism—as concepts that are both separate and overlapping—can do together. I explain what I mean by slow crisis and offer a history of the term, I use queer temporality as a guiding theoretical concept for demonstrating the possibilities in imagining different temporal approaches to crisis, and I show how an apparent feminist understanding of efficiency can utilize queer temporality to operationalize slow crisis as a concept. In sum, this chapter uses slow crisis as a pivot point to show what queer theory and apparent feminisms can and should do in a reciprocal relationship.

Chapter 4 shows where apparent feminism can go and what it can do through detailed analysis of pre- and post-crisis communication related to the Deepwater Horizon Disaster. It offers deep context for the disaster itself—an important step since rhetorical feminisms like apparent feminism must understand context to be appropriately, rhetorically reactive. Building on chapter 4, chapter 5 opens with a mapping of responsibilities and then adds an additional layer of analysis by demonstrating how transcultural and intersectional approaches are absolutely vital to apparent feminist critique.

Finally, chapter 6 offers more applications of apparent feminisms through histories, demonstrating how far we've come, how very far we have to go, and how apparent feminisms can help. This concluding chapter—in defiance of what concluding chapters are supposed to do—offers pathways without solutions, ideas without endings, and a refocusing on bodies and notable people toward theorizing new efficiency frames. That is, it leaves readers with a new way of thinking without telling them what to think. This final chapter keeps an eye to foreshadowing additional applications for this original methodology and reflects on the potential for this methodological approach to create openings for further theorizations of feminist technical communication.

2
APPARENT FEMINISMS

Toward the end of my graduate studies, I taught a class on rhetoric that used feminisms as an organizing feature. I asked students to view the many "traditional" contents of the class through a feminist lens. While most students were open to this and did well, it is often the moments of resistance that I tend to remember, perhaps because they taught me so much. Two such moments have been formative for me. The first was from a student who I'll call Daryll. Daryll was fascinated with my approach and with learning about the Greeks. He was interested, invested, and persuaded. In class, we discussed how no extant works exist from either Socrates or Aspasia, yet one is well-known and commonly accepted as having been a real person while the other is not. During class discussions, Daryll often referred to Aspasia and wondered aloud whether she had been real, lamenting that we will never know since we don't have anything she wrote. A Black man, he sometimes drew parallels to Black historical figures whose work has similarly not been granted its appropriate place in history. At the same time, he often took Socrates's reality as fact. I want to emphasize that Daryll was a receptive, persuadable audience. When other students or I corrected him, he apologized for thinking of Socrates and Aspasia differently and corrected himself. And yet, on Daryll's final exam, he wrote that we only know about Aspasia *because Socrates wrote about her.* This erasure of a woman, and the assumed reality of a similarly situated man, from such a persuadable audience—one with whom I had discussed the parallel situations of Aspasia and Socrates multiple times—surprised and frustrated me. My reaction was undoubtedly in part that of a novice teacher, but this was a moment that really drove home for me how deeply internalized misogyny can be for all of us, even as we are actively trying to do and be better.

I begin this chapter by making explicit some of the organizing features of my thoughts as I developed apparent feminisms. The course I described above was formative for me in thinking about a new theoretical approach that could meet an enthusiastic student like Daryll where he was, support the feminists and women in the course, and better redirect the attentions of students who demonstrated resistance to feminist ideas. I mean for apparent feminisms to be temporary, malleable, and permeable; this includes being useful across multiple disciplines. For this methodology to be as flexible as I hope, I believe I need to explain where it came from and offer its potential users enough agency to chart new trajectories for the future. In other words, I believe it is useful to know the full, rich context from which this methodology emerged.

Around the same time as the class I described above, the Deepwater Horizon Disaster (DHD) became a topic in world media. I was in the second year of a PhD program, and my frames for understanding this event as it occurred were rooted in Certeauian logic and were based on place/space and ecology/economy. While these ideas were central to my thinking—which I explain in more detail later—I found myself piecing together a number of other concepts and methodologies to make sense of the metaphorical and literal ripples emanating from this and concurrent events. Those concurrent events mostly revolved around reproductive rights and the legal rhetorics that constrain and protect them. These years (roughly 2009–2013) marked a peak in state-level anti-choice legislation and attempts to instantiate such legislation. I had a sense that connections existed between the DHD and anti-choice legislation, but it took the development of apparent feminisms to help me contextualize the rhetorical assemblages at parallel work in both situations. Eventually, as my interest in the Deepwater Horizon oil spill and its effects and narrations and my interest in rhetorics of reproductive justice continued, I worked to shape apparent feminisms as a flexible frame to help myself make sense of what I was observing and, more important, to help technical communicators (re)view the world and guide their actions.

This chapter offers a detailed explanation of apparent feminisms as a methodology. While apparent feminisms grew out of the situations I described above and did so at a time when anti-choice legislation was on the rise, these orientations are not limited to addressing situations having to do with reproductive rhetorics, natural disasters, or pedagogy. In fact, one of my main goals with this text is to demonstrate not just that apparent feminisms can intervene in situations where their necessity might be less obvious but that that very lack of obviousness is often a sign of the necessity of intervention. Thus, through this book, I demonstrate how apparent feminisms developed out of concern for reproductive justice, where such intervention is perhaps expected, but also how they can be applied in a variety of contexts that may seem far removed from that origin—namely, the Deepwater Horizon Disaster. This chapter takes up reproductive justice as a way to show where apparent feminisms came from. Chapters 3 and 4 show where apparent feminisms might extend. To be clear, I am not claiming that apparent feminisms are universally applicable—in fact, I have concerns about some contexts in which this methodology might be applied (which I will touch on later). However, this book emphasizes the complementary work social justice theories and apparent feminisms can do together—indeed, I emphasize that they cannot be separated.

In this chapter, I offer context for the development of this theoretical approach and then describe the design of apparent feminisms as a methodology, including explaining its audiences and major goals. I then move into discussing and building on the existing technical communication research most relevant to this project; this includes feminist technical communication work as well as work in technical communication that lays the foundation for humanistic and social justice inquiries. An extension of the literature review of feminist technical communication scholarship in chapter 1, this section is aimed at pointing out common threads toward establishing both points of strengths and critical gaps. Then, I discuss the interdisciplinary scholar-activists whose work apparent feminism also builds on (primarily Bray 1997; Halberstam 2002, 2005; hooks 1989, 1994, 1995, 2004; LaDuke 1999, 2005; Mohanty 1988, 2003) and speaks back to, as well as the disciplinary contexts in which this theory seeks to intervene. These scholars help me think more broadly about the applicability of apparent feminist approaches and about the reach of technical communication artifacts, and they help fill in fissures and reinforce stress points in existing feminist technical communication literature. Further, they help with bridging divides toward a transmigratory feminist agenda precisely because they reside outside what we might recognize as the discipline of technical communication. Finally, I orient this new approach as a call to action for technical communicators.

ORIGINS

Public discourse about reproduction—just one of several possible examples of contemporary public discourse about feminist concerns—provides an urgent need for work at the intersections of feminisms and technical communication. Visible rhetorics,[1] medical rhetorics, and rhetorics of science and technology in particular have often been utilized in service of radical legislation that points to a serious need for feminist perspectives on legal technical communication occurring in the public sector.

For example, several states have passed or proposed legislation requiring people seeking abortions to undergo various medically extraneous procedures. A major push for such legislation peaked around 2012.[2] In February of that year, Texas senator Dan Patrick (R–Houston) won the support of a district judge, allowing a new law (formerly HB 15) to amend the 2003 Woman's Right to Know Act (also known as chapter 171 of the Texas Health and Safety Code) so that people[3] seeking abortions must have an ultrasound, hear the doctor describe the "child," see a sonogram or listen to a heartbeat, and then wait twenty-four hours before beginning the procedure (Arizona Women's Health and Safety Act 2012; Eckholm and Severson 2012; Jones 2012; SB484 2012; Texas Health and Safety Code 2012). Feminist responses to such laws were swift; Dahlia Lithwick (2012, para. 1), for example, wrote that a similar law requires that "women seeking an abortion in Virginia will be forcibly penetrated [by being required to have a transvaginal ultrasound] for no medical reason . . . [which] would constitute rape under the federal definition." As it often does, backlash to this feminist outcry asserted a deep and malicious feminist bias.

This situation left me, a rhetorician and technical communicator, with a series of questions: how might feminist technical communicators persuasively point out the bias inherent in all worldviews, even those that people often perceive as neutral? How might they intervene in unjust situations, particularly in technical contexts in which objectivity is highly valued? And how might they best decide which situations are most deserving of this sort of attention? How might I develop a rhetorical feminism to help stakeholders wrestle with these ideas? From these questions, I built an apparent feminist methodology that highlights the urgency of responding to such contemporary, oppressive situations. Technical communication scholars and practitioners need an apparent feminist methodology to intervene in situations in which technical documentation unfairly and uncritically engages in oppression while feigning objectivity.

To bolster the continued importance and success of technical communication as a discipline, we must pay attention to the powerful ways critical and cultural scholarship such as apparent feminisms can help us reshape our world and the ways we move through it. For example, the aforementioned legislation evidences that female bodies are under attack, and apparent feminist methodologies can point to methods by which technical communicators can intervene. For example, countersuits to laws that restrict abortion access were filed in a number of states. In addition, apparent feminisms can be used to leverage new understandings of gendered relations in business, education, and social structures.

Apparent feminisms are methodologies that seek to recognize and make apparent the urgent and sometimes hidden exigencies for feminist critique of contemporary technical rhetorics. This approach encourages a response to social justice exigencies, invites participation from allies who do not explicitly identify as feminist but do work that complements feminist goals, and makes apparent the ways efficient work actually depends on the existence and input of diverse audiences. Thus, this methodology recognizes a diversity of credible (and biased) perspectives while valuing embodiment and interrogating the way bodies—especially female bodies—are represented. An apparent feminist methodology is necessary because of (a) current political trends that render misogyny unapparent at the nexus of social, ethical, political, and practical technical communication domains (Hart-Davidson 2001; Johnson 1998; Miller 1989); (b) the ubiquity and globalized, networked distribution of uncritically negative responses to the term *feminism*; and (c) a decline in centralized feminist work in technical communication since around 2005.

DESIGN AND GOALS OF APPARENT FEMINISMS

I envision apparent feminisms as temporary, permeable, and flexible. In the interest of transparency, I had become partly convinced that apparent feminisms had outlived their usefulness a few years ago, when reproductive rights seemed to be gaining ground and more technical communicators were talking about social justice. When the Association of Teachers of Technical Writing first announced the theme for its 2019 conference, I thought social justice had arrived in a big way and apparent feminisms could and would fade away under this umbrella. More recent exigencies along with a group of brilliant women scholars from North Carolina State University (mentored by Stacey Pigg) persuaded me that apparent feminisms remain urgent. Thus, I have returned to the foundations of this theoretical approach even while recognizing—and cautioning readers—that it is open to interpretation, remixing, extension, and, someday, it is hoped, obsolescence.

The term *apparent feminism*, like the dynamic spirit of rhetoric, is always in flux; it can be understood in different ways given different contexts, and it encourages practitioners to reflexively understand that any given apparent feminist approach is only one possible understanding of any given situation. I chose this term because the word *apparent*

- Signifies the need to make feminist issues in everyday life conspicuous.
- Emphasizes the valuing of explicitly feminist perspectives on such issues. Calling attention to a site for intervention is not enough; we must also point to the value of feminism as a response.
- Points to the lack of apparency of important terms and their definitions in particular rhetorical contexts. For example, because *efficiency* is such a valued term in technical communication, it too often goes undefined and unexamined.
- Calls on the notion of presence without using ableist language. As Christa Teston (2012, 205) put it, "Presence is about more than being visible to the eye. Presence is about being visible to one's consciousness." That is, the term apparent not only draws on the idea of constant presence but is also an active term that avoids rendering non-seeing bodies as abnormal.
- Responds to the uninformed ways in which feminism is often criticized. The notion of apparency, in its incarnation as denoting something potentially false, calls subversive attention to knee-jerk attacks on feminism.

Readers may notice that I usually pluralize the term; this is to recognize that the methodology itself can be enacted in multiple ways and, further, that it might (I hope) be enacted in ways I am unable to imagine because of my positionality. I ask users of this methodology to consider multiple meanings of the term *apparent* to "marshal the knowledge necessary" to resist the ideology of the term as used in this context "while at the same time also speaking in, and from within" that context (Sandoval 2000, 44). Apparent feminists can draw on Chela Sandoval's theory of oppositional consciousness as a "politically effective means for transforming dominant power relations" (44) in their use of the term *apparent* to describe their feminism.

In part because of this flexibility, all technical communication scholars and practitioners can benefit from taking up an apparent feminist methodology. (In fact, I believe apparent feminist approaches can be applied much more widely, though my interest in this book is mostly to explicate their use in the particular context of technical communication.) Our main goal as public intellectuals and technical communicators should always be to do good work in the world; in this case, the meaning of good is based in part on our audiences, readers, and users. Apparent feminism can help us think about the diversity of potential audiences, what productive or efficient work is, and how to do our work in socially just ways.[4] Scholars and practitioners of technical communication can use apparent feminisms to think beyond the usual or initial audiences for any given piece of communication, and this orientation can remind us to have reflexive conversations about what constitutes good, productive, efficient (and so on) work. Further, apparent feminism's flexibility invites both feminist and non-feminist technical communicators to the table. A technical communication scholar need not self-identify as feminist to use this methodology;[5] apparent feminism purposefully does not demand such an identification. Rather, it seeks to highlight the opportunities for productive coalitions among many kinds of technical communicators.

The following subsections elucidate the three major goals of apparent feminism: making more apparent the need for feminist interventions, hailing non-feminists as allies, and demystifying the relationship between feminism and efficiency. These goals are interrelated, and I separate them here only to convey their unique attributes. I conclude with a reassembled iteration of these goals.

Making More Apparent the Need for Feminist Interventions

Activist-actress Laverne Cox has been widely quoted as saying, "It is revolutionary for any trans person to choose to be seen and visible in a world that tells us we should not exist." Indeed, visibility is often in and of itself a revolutionary position. This fact provides a starting point for understanding the goals and interventionary strategies of apparent feminisms. First, apparent feminisms recognize the urgent exigencies for feminist critiques of contemporary politics. Politics here is broadly construed, encompassing any action or policy aimed at accruing or maintaining power. Of special interest are those political artifacts that masquerade as something else. Specifically, technical documentation is often about power while appearing in a guise of objectivity.[6] Apparent feminisms recognize the importance of making a feminist philosophical perspective apparent in such critiques. My drive toward such apparency results from the range of responses I have encountered in talking about feminisms with students, friends, family members, and colleagues. I have found that responses to feminisms are often tied up in rhetoric about bias.

While teaching an upper-level undergraduate seminar on feminist rhetorics (the same class I opened this chapter with) at Illinois State University, I encountered many situations in which students wanted to talk about bias. Most students readily accepted the validity of a social constructionist view of the world, and most were willing to agree that all perspectives, all worldviews, contain some degree of bias. Nevertheless, "bias" persisted as a critique of feminist perspectives throughout the course. One student left a single comment on the end-of-semester evaluation that I will never forget: "Feminism is a biased perspective." I remember covering my face with my hands in frustration when I read this, not because the statement is incorrect but because its implication is that other, unbiased perspectives exist. The smart, mostly well-intentioned students in this course seem to me accurate representatives of this trend in larger contexts. That is, people seem to believe that feminism is a *particularly* biased perspective. This belief in feminist bias—and, by extension, an implied belief in relative objectivity—left me wanting to explore the possibilities of rhetorical, reflexive feminist approaches to technical communication.

Consider, for example, Texas's 2012 amendment to the 2003 Woman's Right to Know Act. This amendment—under the guise of a "woman's right"—requires women to have a medically extraneous ultrasound procedure performed before being able to access abortion services. Feminist responses to this law have often been attacked as "biased" and "radical," although they are surely no more biased or radical than the legislation itself. The presumption of bias is not an issue limited to feminism. The field of rhetoric—and, by extension, technical communication—struggles with the notion of bias as it interfaces with perception; this struggle enhances the exigence for an apparent feminist methodology that can help point to the universality of bias. Consider, for instance, that the main criticism of Lloyd F. Bitzer's (1968) classic "Rhetorical Situation" is his reliance on a realist or objectivist theoretical approach. Richard E. Vatz (1968, 154) argued that Bitzer's realist philosophy has "important and, I believe, unfortunate implications for rhetoric," which Vatz articulates as the difference between defining rhetoric as a creative act or a simple observation. In other words, Vatz argued that Bitzer leaves out the important role of human perception and discursive interpretations of perception in the rhetorical situation. In reinvesting the rhetorical situation with human perception, Vatz reversed each of Bitzer's main conclusions: "For example: I would not say 'rhetoric is situational,' but situations are rhetorical; not '. . . exigence strongly invites utterance,' but utterance strongly invites exigence" (159). Indeed, when considering the urgent need for apparent feminism, we see that the situation is highly rhetorical and that particular kinds of utterances, texts, and rhetorical acts create public biases—or, we might say, crises—which, in turn, demand an apparent feminist response.

The urgency for this methodology arises from the fact that feminists who attempt to respond to attacks on female bodies that are perpetrated through technical documentation (as in the Texas act just cited) and technology (in this example, the ultrasound equipment) are often written off as hysterical or hostile—or are altogether silenced. Many scholars whose work qualifies as apparent feminism rebel against this silencing and simultaneously against the separation of embodiment and technological culture (Munster 2006; Nakamura 2008). Anna Munster (2006, 9, 3) argued that digital histories "have limited the present and future possibilities for bodies in culture" and that, in response, we should create an alternative digital genealogy that takes "body, sensation, movement and conditions such as place and duration into account." Munster called for "a more expansive conception of materiality" that "can help us to draw intensive connections between the actions and affects of bodies and forces of digital code" (85). As many scholars have noted, technologies and bodies are codependent and always acting upon each other (Bray 1997; Frost and Haas 2017; Nakamura 2008). Apparent feminism is concerned with the effects of technologies (e.g., those included in abortion, ultrasound, medicine) on bodies and with who gets to decide when the two interact in particular ways.

Working toward feminist apparency and intervention can call attention to different understandings of such technological processes that shape cultural commonplaces. Apparent feminist projects might include calling attention to highly visible rhetorics—such as legal documents—that shape our ideologies. The cultural importance of these documents is already readily apparent, but such technical artifacts are not often considered to be sites of culture making. Apparent feminism seeks to recover understandings of such rhetorics as culturally saturated, and it seeks to question who stands to benefit from such culturally saturated spaces. To explain, pointing out that an ultrasound exam is a medically extraneous procedure in relation to abortion services is an example of how an apparent feminist might make more apparent the ideologies behind such abortion rhetorics.

A push for apparency as described here may—or should—give rise to conversations about access. Access is a feminist issue and not a simple one. The notion of apparency might suggest that a feminist stance means always advocating accessibility no matter the context. This stance is more akin to notions of transparency, which suggest a polar position. Apparency, in contrast, might lead us to think about a more nuanced approach. To explain, we sometimes need to weigh access against privacy, safety, or both. As one relatively benign example, I received an email in April 2020—in the midst of a national push for developing self-isolation practices to slow the spread of Covid-19—in which a Women's Center director shared a link to an incredible compilation of articles that put Covid-19 and gender into conversation *and asked that it not be shared on social media.* By circulating this list only among peer networks, the organizer avoided the trolling that would inevitably have accompanied more public distribution and thus avoided putting themself and others through additional trauma during an already traumatic event. In other words, decreasing access while increasing feminist apparency was this person's chosen route to help others while maintaining the safety of their staff and colleagues.

Another important conversation related to the notion of apparency is that of anecdote, a term that recalls a variety of contexts. When I worked as a reporter, anecdote was often a useful way into a larger human interest story. When I took my first methods course, describing something as anecdote was a way of dismissing the value of a research contribution. When I began to study feminisms in earnest, I learned that anecdote occupies a liminal (sometimes awkward) position somewhere between "the personal is political" and notions of academic rigor (enforced enthusiastically by many academics in evaluating others' work).

Anecdote is important for apparent feminisms because it is sometimes the best or even the only way to make a feminist urgency evident. In her 2002 text *Anecdotal Theory,* Jane Gallop attempts to reunite anecdote with theory, suggesting that they do not have to exist in opposition. Citing Barbara Christian's (1987) "The Race for Theory," she argues for an alternative way of theorizing that is situated in experience—in anecdote. This focus on experience has important implications for feminists, who have long argued that those with marginalized identities are implicitly left out of much of what passes for "theory" or "empiricism." Reinventing experience as (at least a contributor to) theory opens up gateways for those who have been left out to enter. Gallop and Christian are far from alone in this decolonial call to pay attention to story, anecdote, experience (Blackmon 2007; Haas 2012; Jeyaraj 2004; Jones, Moore, and Walton 2016; King 2008; Williams 1991). Gallop's (2002, 2) take on the relationship between anecdote and theory is fairly narrow: "During the nineties I experimented with writing in which I would recount an anecdote and then attempt to 'read' that account for the theoretical insights it afforded. It is this particular practice of theorizing that I want to indicate by the title Anecdotal Theory." Thus, it is less her method than the spirit of it that I seize on here: "I have wanted to seize the fallout of event; that is the point of anecdotal theory.

I hope the anecdotal can wedge open my own theoretically predictable discourse. Above all, through the anecdotal, I want to leave open the chance for something to happen" (157). Gallop is looking for the fissures in theory, the places that experience and context can lay bare but that theory misses.

This urgency for feminist intervention also exists in disciplinary contexts, which, in turn, affect public understandings. Feminist work in technical communication began with the recognition of gender as a social category, as shown in the 1991 issue of the *Journal of Business and Technical Communication* that focused on gender as part of a cultural turn. The 1994 issue of *Technical Communication Quarterly* then focused on women's contributions to technical communication, a technique that has sometimes been critiqued as the add-women-and-stir approach. Between 1994 and a historiographical push in 1997, some feminist technical communicators recognized that "stirring" women into the field was entirely reshaping its content (Durack 1997). These knowledges and histories have expanded the scope and expertise of the field and made more apparent how insidious technical communication can be in the construction of everyday life for all people, including but not limited to women. The realization that feminist influences can produce new and robust sets of perspectives on disciplinary content areas should inspire more technical communication scholars to take up feminist perspectives in their work.

In review, apparent feminism is a methodological approach that can work within the field of technical communication by emphasizing the importance of making feminisms explicit. It responds to the suggestion of a post-feminist world by reasserting the relevance of and need for feminist perspectives in contemporary social politics. Apparent feminisms support:

- Ways of pointing out and resisting oppressive technical communication and rhetoric, particularly when such oppressions are based on sex and gender
- Our understandings of objects, processes, technologies, and ideas often considered to be neutral or objective as being culturally saturated
- Persuasive responses to detractors of feminism
- Discussions about identity politics and the problems of associating oneself with the term *feminism*.

To be sure, several of these concerns also bear on the other interrelated goals of apparent feminism. I will make those connections more apparent as follows.

Hailing Non-feminists as Allies

The recognition of allies is an important rhetorical strategy for apparent feminisms. That is, apparent feminist technical communicators are interested in building coalition with those who might appear to be feminist in their activism or ideological perspectives but do not embrace that label—often for good reason. As one example, Lorgia García Peña (2022, 55) writes of her elders—women who sparked her own rebellion, defended her natural hair, and took down abusive husbands through coalition: "Perhaps they would not have called themselves feminists—a label that in the Dominican Republic, for the entire twentieth century, was associated either with formal political party organizing or with 'foreign' white women. Their feminism was simpler, quieter, and more tacit. It revolved around mutual care for each woman in the community through daily practices of support and presence."

Clearly, these women were engaged in work a feminist technical communication scholar would celebrate, even if the label *feminist* didn't apply. As a more in-field example, an apparent feminist scholar of technical communication and a non-feminist-identified lawyer interested in working to overturn unjust regulations (e.g., the Texas Woman's Right to Know Act) might form a symbiotic partnership, to great result. This advocacy of coalition performs more than one function. First, it creates more spaces for conversations about the value and challenges of feminist identification—thus potentially also increasing public exposure to feminist rhetorical values. Second, it provides an alternative to the exhausting, decades-old notion that feminists themselves must bear an increased ethical burden to demystify and fight against gender inequities.

This invitation to allies comes with the necessity of giving up some measure of control to gain new, interdisciplinary, diverse approaches to technical communication. In this way, apparent feminism encourages increased inclusivity and diversity in our understandings of and conversations about gendered systems. Apparent feminism requires not just that feminists take a stand—as they often have done—but that they allow space for those who do not self-identify as feminists to enter the conversation. It allows a person to do feminist work and even to recognize it as such if they wish without necessarily self-identifying as a feminist. In other words, apparent feminism recognizes that it is itself a methodology with particular investments (or biases) and that other (equally biased and equally legitimate) methodologies might be complementary. For instance, apparent feminism lends itself to a variety of cultural studies approaches that have been taken on by technical communication scholars. It can contribute to and dialogue with work such as Jeffrey T. Grabill's (2007) scholarship on citizenship practices and Angela M. Haas and her coauthors' (2002) research on girls' technological literacy, for example. Apparent feminisms, therefore, can sponsor bridges between a wide variety of both interdisciplinary and technical communication scholar-practitioners, including those invested in critical race theory, gender theory, postcolonial and decolonial theory, cultural theory, and more.

In the time since I originally proposed apparent feminism as an approach for technical communication (Frost 2013a), some scholars have begun to make distinctions between being an ally and being an accomplice. This distinction is important and worth drawing out here. According to Laura Gonzales and Haas (2019), accomplice-ship is more active than ally-ship. An accomplice will intervene; an accomplice will put their body on the line; an accomplice can be counted on to do the right thing even when they get no credit. Gonzales and Haas make an important distinction and a persuasive argument. I continue to use the language of allies here for one simple reason: apparent feminism is asking us to seek out people who aren't yet active as feminists—people who are, by definition, passive toward or suspicious of feminist causes. In other words, apparent feminism seeks out allies with the possibility of turning them into accomplices. The aim of hailing allies does not prevent us from also seeking accomplices. Seeking accomplices is not the main thrust here because the hope is that those accomplices are already part of the work that's happening.

The above is not to say that ally-ship is uncomplicated. Sara DeTurk (2011, 584) found in her study of how allies challenge social injustice that "several participants resisted [DeTurk's] initial definition of 'ally.' " Often, this resistance had to do with the definition of ally as someone with relative privilege; the participants in her study saw themselves as able to perform ally-ship for people with equal or more privilege than they had themselves. Along the same lines of attention to power, DeTurk notes that ally-ship often reflects, mimics, and occurs within dominant power structures; however, she also finds that allies make informed tactical decisions to perform their work. In other words, acts of ally-ship occur in contexts of power incongruities by definition and therefore are constrained by those inequities. This means apparent feminists must run risks if they are to enact change. As with apparency, awareness of risk and consideration of the effects of the ways identity characteristics might amplify risk or (alternately) garner privilege are paramount.

Much of the above has to do with the term *ally* as a label. Apparent feminists should also think about allies (and ourselves) as subjects. Louis

Althusser (2006, 105) tells us that "ideology hails or interpellates individuals as subjects." Apparent feminisms are about moving beyond those subjects of interest who are most easily or obviously recognized; apparent feminists must work to hail allies who may not be immediately obvious or recognizable—those who are not already feminist and those we would not ever expect to be feminist. Althusser further argues that nothing can take place outside ideology and, indeed, that becoming conscious of ideology is difficult; "ideology has always-already interpellated individuals as subjects" (106). Thus, hailing non-feminist allies may seem on its face to be a simple maneuver. However, doing so in a way that demonstrates a sincere commitment to dialog is actually both rhetorically and internally challenging, and doing so without losing oneself is even more so. In revisiting her 2003 article "Rhetoric on the Edge of Cunning; or, the Performance of Neutrality (Re)Considered as a Composition Pedagogy for Student Resistance," Karen Kopelson (2020, 14) writes about "my rather quick slide away from the performance of neutrality and toward greater ideological transparency in my classrooms" and clarifies that her true point is about being responsive to context—responsive to the rhetorical situation. The rhetorical situation, of course, includes the feminist/rhetor's body; and "the co-optation of traditional academic postures such as authority, objectivity, or neutrality by marginalized teacher subjects worked to disrupt students' identity-based expectations and oppressive stereotypes" (15–16). Specifically, Kopelson worries over a situation in which students at the end of a course did not identify her as lesbian and how this would mean that queer students might think they had never had a queer teacher and therefore would not see themselves as potential teachers or leaders. In this political era, this sort of "self-erasure" is something she finds especially troubling. Indeed, it could mean that potential allies and potential future accomplices have trouble identifying one another.

Of great influence in taking up the term *allies* is a history of gay, lesbian, bisexual, transgender, transsexual, and queer activism that has both developed and reflexively critiqued ally-ship (Carlson et al. 2019; GLAAD 2020; Kolasinksi 2017). The term has also been used by Indigenous activists and social justice–oriented groups (Heaslip 2014). Because ally-ship draws on these rhetorical histories, it offers affordances for understanding some of the ways apparent feminists might engage in persuasive practice. As one example, "straight allies" often discuss the importance of community membership; that is, these allies might themselves identify as straight but also consider themselves part of the LGBT community—and, importantly, many LGBT-identified members of the

LGBT community consider them valuable members as well. Apparent feminists can follow this model by inviting allies into their communities.

Hailing allies is especially important in the era of the "echo chamber" (Gurak 2018). Social media, which purports to connect us in ways we have never been connected before, also allows us to filter what we encounter of others' rhetoric in ways we have never been able to before. As more than one popular source has noted, this can have the result of producing a digital echo chamber where one's own views are continually reinforced. For example, the popular *Wall Street Journal* project "Blue Feed, Red Feed" shows what two different social media account users' feeds might look like, one with traditionally liberal and one with traditionally conservative perspectives (Keegan 2019). While social media platforms are powerful tools for connection, this particular usage runs counter to apparent feminist thinking. Hailing allies means inviting people to the table who have views different from our own—not only in hopes of persuading them of our views but in an honest effort to understand their perspectives as well.

Marking work that complements feminist goals is a historical as well as a contemporary project. That is, apparent feminism might help us find allies whose work has already been done. This notion of apparency after the fact is particularly important because it can help recover feminist works that were not previously recognized as such. For example, although some of the works I discuss in this chapter are not apparently feminist, I was able to use them in ways that are explicitly feminist without violating the spirit of their messages; indeed, I hope my work here—and my recognition of the apparent feminism of other works—will further the projects of scholars like Francesca Bray, Jack Halberstam, bell hooks, and Winona LaDuke—whose work I detail later in this chapter as foundational to my thinking on apparent feminisms. This recognition of feminist works in places where they were not previously apparent as such can build on the work of rhetorical feminists like Rachel Chapman Daugherty (2020, para. 5), who, in turn, builds on prior feminist thinking: "Feminist rhetorical historiographers like Jessica Enoch have established the 'construction of memory' as a rhetorical act ('Releasing'), tracing the ways that ideology shapes the inclusion and exclusion of memories from archives and public memory. I apply Enoch's concept of construction to the archival submission and selection criteria of the Sister March archives, treating archival construction as a rhetorical site for inclusion through representation in public memory. As a rhetorical analytic, archival construction encourages researchers to identify how, why, and for whom public memory is made." Thus, this (re)construction of memory, of feminist apparency or lack thereof, is an act built on iterative feminist influences.

Finally, because apparent feminism is a methodology based in technical communication, part of its purpose is to value the expertise of all of our potential allies. To explain, the term *technical* refers to the existence of a specific, specialized audience. While specialization is potentially exclusionary, apparent feminism makes clear that all individuals have their own diverse technical expertise. Technical communication practitioners are well prepared to recognize, honor, and work in concert with these diverse skill sets. We are a discipline very used to working with subject-matter experts. How we construct the social frameworks that recognize expertise—how we invite peers, including non-feminist allies, to the conversation—is what really matters.

Demystifying the Relationship between Feminism and Efficiency

Apparent feminism resists refusals to examine our disciplinary terms, instead requiring that we interrogate the invisible underpinnings of these terms to make their values more apparent. At this moment, one of the most important terms requiring apparent feminist attention is *efficiency*. A particular understanding of the term *efficiency* has become rather naturalized for technical communicators; thus, contextual definitions of efficiency too often go unarticulated. Apparent feminism points to those unarticulated parts. Specifically, apparent feminism responds to the stifling of cultural diversity in the name of efficiency, which is a term that—contrary to some understandings of it—relies on the existence of diversity for its value (for a detailed critique of modern usages of the term *diversity*, see Ahmed 2012). To explain, efficiency is commonly understood as the balancing point at which we achieve the best result while expending the least amount of energy. However, technical communication as a discipline is currently most focused on the latter part of that definition. Apparent feminist technical communicators must rearticulate efficiency as focused primarily on audiences as a component of best results—and must draw attention to which audiences we focus on.

In advocating this new focus, I also recognize that the balance between these two criteria for determining efficiency is contextual. I believe technical communicators, in general, need to shift our focus from energy expended to results produced because of troubling patterns within disciplinary rhetorics. However, apparent feminists must always keep in mind that the best balance to achieve ethical, effective, socially just technical communication will be different depending on the rhetorical situation.

Apparent feminism seeks to make the symbiotic relationship among feminism, diversity, and efficiency more apparent. Consider the connection between diversity and feminism. As we know, feminists are concerned with people other than just those who self-identify as women (Butler 1990; Evans 2012). Social justice—which focuses on justice for the wider community—is the ultimate goal of apparent feminism, so apparent feminists are interested in being inclusive of people marginalized by hegemonic understandings of ethnicity, class, and ability in technical communication theory and practice. In essence, when we invoke the term *efficiency*, we must also make apparent the contextual implications for ethics. For example, Sam Dragga and Dan Voss (2001, 267) discussed a statistician who, "prizing efficiency and ignoring ethics," insisted on using only scientific data. This phrasing makes apparent an understanding of efficiency as distinct from ethics, an understanding I believe is too common among technical communicators and one we have an obligation to disrupt.

The notion of efficiency relies on the existence of diversity for its value;[7] apparent feminism argues that being inclusive of cultural diversity contributes to the potential for efficiency. This reasoning might initially seem counterintuitive to some technical communicators, who may argue that a traditional style exists and that including more audiences means doing more work and would therefore contribute to inefficiency. However, because technical communication is consumed and produced by culturally diverse audiences and because technical communicators cannot always predict the rhetorical velocity (Ridolfo and DeVoss 2009) of their texts, producing texts that can be usable and useful to as many potential audience members as possible in a given context is necessary in our increasingly globalized and digitized discipline and world (Nakamura 2008). Remembering that efficiency is defined as the balancing point at which we achieve the best result from the least amount of energy, we must recognize that this increased audience represents a significantly greater result. Ultimately, then, apparent feminist practices have the potential to produce more efficient work—if we operate under this reimagined version of the term *efficiency*. Consider Carolyn Jones's (2012) account of her interaction with the Texas Woman's Right to Know Act. After Jones and her husband learned that their unborn son would face a lifetime of pain and suffering, they decided that terminating the pregnancy was the best path forward. After a counselor explained the process then required by the state of Texas, Jones wrote:

> "I don't *want* to have to do this at all," I told her. "I'm doing this to prevent my baby's suffering. I don't *want* another sonogram when I've already had two today. I don't *want* to hear a description of the life I'm about to end. Please," I said, "I can't take any more pain." I confess that I don't know why I said that. I knew it was fait accompli. The counselor could no more change the government requirement than I could. Yet here was a superfluous layer of torment piled upon an already horrific day, and I wanted this woman to know it. (para. 17, original emphasis)

While a certain percentage of the population might consider the law to be good or productive, Jones is certainly not a member of that group. If her positionality had been taken into account, it might have produced a more socially just—for a wider potential audience—set of regulatory rhetorics.

Current understandings of the term *efficiency* in technical communication constrain the goals of apparent feminism, and this constraint is of utmost importance because the term has such value to the discipline. That is, with such value placed on efficiency, adhering to whatever understanding exists for that term in a particular cultural context becomes of utmost importance because anyone who can be labeled inefficient is a threat to the status quo. Efficiency can easily become so embedded as a cultural value that it is no longer explicitly discussed—the shifting balance of energy expended versus goodness done is not articulated—and it is then a small step to using efficiency to justify racism, sexism, ableism, and other evils. Robert McRuer (2006, 88) pointed out that the work of efficiency experts "played a large part in the emergence of the identity of the able-bodied worker." In determining what was efficient, these experts had to construct ideas about which bodies could work most effectively, therefore privileging some bodies over others.

We can see new strains of this ethic of efficiency in many modern conversations; one such conversation that is particularly important for technical communicators is about digital education and the accessibility of technology. Kevin Eric DePew, T. A. Fishman, Julia E. Romberger, and Bridget Fahey Ruetenik (2006, 64) suggested that competing ideologies of efficiency in digital education require our attention and that "the privileging of efficiency over pedagogical principles of dialogism, collaboration, and an emphasis on process must be resisted." In a similar vein, Virginia Eubanks (2011, xv) warned that social justice advocates who emphasize technology accessibility without accompanying rhetorical training "as the route to greater prosperity and equality for all Americans" are engaging in "a familiar but dangerously underexamined species of magical thinking." In other words, the common belief that more technology is an equalizer and a road to greater efficiency for large audiences leaves out a lot of ethical complexities.

Because a particular understanding of the term *efficiency* has become so naturalized for technical communicators, parts of its definition often go unarticulated. Apparent feminism addresses this by making apparent those unarticulated parts. To illustrate, one fairly recent definition of efficiency is "the rate at which users can successfully complete a task after they have learned the system" (Simmons and Zoetewey 2012, 266). In this case, efficiency is defined as being almost entirely about time or energy spent. Apparent feminists point out that the aforementioned definition could be improved upon by talking about the quality of the task completed. Apparent feminists might revise this definition to focus on the importance of multiple identities of users. For example, W. Michele Simmons and Meredith W. Zoetewey ultimately criticized the traditional notion that engaging the multiple identities of a single user is a drain on efficiency. They made apparent that they privilege these diverse identities and the qualities they bring to the situation, thereby helping to dismantle the efficiency-diversity dissonance.

In shifting the focus of efficiency away from energy and toward people, apparent feminist technical communicators must constantly ask themselves if their work is efficient in multiple long-term and short-term frames of reference. For example, is a particular action efficient in terms of serving both their employer and the interests of social justice? Cezar M. Ornatowski's (1992) approach is one I adopt as apparent feminist work because he made it apparent that the particular actors involved in a rhetorical situation determine the way efficiency is enacted. Ornatowski pointed out that a technical communicator is always held responsible for two contradictory goals: they are supposed to be objective and yet serve the interests of their employer. Using textbooks as the (admittedly limited) basis of his critique, Ornatowski suggested that efficiency is "understood in terms of usefulness to employers" (93) and that technical communication that does not serve the employer is deemed inefficient. He resolved this contradiction with the argument that efficiency is always political (100). For apparent feminists, understanding efficiency as political encourages caution—but not necessarily resistance.

Apparent feminism suggests that we can retain efficiency as a primary guidepost for determining the value of technical communication as long as we are careful in articulating our understandings of what efficiency means. Steven B. Katz's (1992) work with the term *expediency* provides an example of technical communication's use of efficiency and the ways it moves through our rhetoric without requiring us to critically analyze its use. Katz—who was especially careful with his terms—repeatedly invoked efficiency, but the term is so familiar to his technical communication audience that he did not define it. To provide some context, Katz criticized expediency as a criterion for determining effective technical communication by beginning with the example of a Nazi memo that is technically effective but ethically repugnant. In the following quotations from Katz's article, the emphases on usages of efficiency rhetoric are mine:

- "The writer shows no concern that the purpose of his memo is the modification of vehicles not only to improve *efficiency*, but also to exterminate people" (257).
- "The whole society organized into a death machine for the *efficient* extirpation of millions, lauded by the Nazis as a hallmark of organization, elegance, *efficiency*, speed, all of which became ends in themselves for those planning and those executing the procedures" (265).
- "Propaganda thus served to create the technological ethos of Nazi consciousness and culture: rationality, *efficiency*, speed, productivity, power" (268).

In all these instances, Katz shows the term *efficiency* being used to emphasize the minimization of energy expended rather than the maximization of positive effects on human lives (which, in this case, would certainly be a negative value). Apparent feminism suggests that efficiency can be used as an ethical end, but only if we understand if as focused primarily on effects on human lives and secondarily on speed and energy.

The rearticulation of efficiency as audience-focused is important because apparent feminism suggests that the disciplinary domination of male perspectives in technical communication—a domination that has been continually reproduced in the name of efficiency—has limited the scope of what technical communication can and should do. When we revise our understandings of audiences as diverse rather than just male, the nature of efficiency changes. This point is especially important in the discipline of technical communication because it is so heavily male-dominated (Allen 1994; Bosley 1994; Flynn et al. 1991; Rauch 2012; Ross 1994; Royal 2005; Thompson and Overman Smith 2006).[8] Understanding efficiency (in its common usage) as something that benefits those already in power, something that reinforces the status quo, something more focused on energy expended than on people affected, prevents us from doing good, ethical, socially just work. But understanding efficiency as something that focuses on both energy expended and people affected, something that places primary importance on the human rights of stakeholders, might lead us toward altogether different practices.

This efficiency as imagined from a feminist perspective is "humanistic, life-oriented, non-exploitative" (Rothschild 1981, 66). If we—apparent feminist technical communicators—are truly interested in doing efficient work, we must embrace diversity to hail larger audiences. Thus, apparent feminists work to make apparent the ways efficiency actually relies on other concepts—like feminism as it supports diversity—for its value. Kellie C. Sharp-Hoskins (2012) uses the rhetorical canons to demonstrate that terms always rely on other terms for their value; by shifting our terministic screens for the term *efficiency*, we can learn to imagine how this term might actually give rise to *other* metrics for assessment. Reimaginings of assessment criteria for what terms constitute efficient (or effective, good, useful, productive—any number of other synonyms that denote an assessment system's values) communication already exist in rhetorically careful investigations of technical communication work. Becca Cammack (2015, 130) asks "whether a connection exists between environmental regulations and our perception of sustainability" and ultimately finds through her case study that transition from "a set of frustrating rules" to "an important contributor toward a global initiative" is possible but that participants may not always see it that way. Rather, participants—probably correctly—intuited the importance of a more comprehensive approach, and "people living outside of the environmental bubble . . . may not see or perceive the same messages in their own exposure" (143). Elena Sperandio (2015), in her investigation of efficiency and customer satisfaction related to technical documentation, advocates seeing technical communication as part of the product.

Another way of thinking about this reframing of efficiency is as an attempt to practice breaking free of the commonplaces that govern how we think about any given topic. In a meta-analysis, then, we might reimagine how we think about and apply feminisms. In other contexts, we could reimagine how we think about ecology and economy to better understand how something like the DHD could happen. Denise Tillery (2018) offers an important first step for doing this work in her book *Commonplaces of Scientific Evidence in Environmental Discourses*, wherein she identifies some common ways we talk about environment and explains how these tropes limit our thinking. For example, she lays out how belief in science is cast as an assumption that we can know and control nature. As another example, she discusses the trope that science is objective and above politics—alongside the trope that science is ideological and deeply impacted by politics, thus demonstrating that presenting such a viewpoint leads naturally to its polar opposite but may elide positions that lie elsewhere on the spectrum or on a different spectrum altogether. In other words, breaking free of commonplaces means not just rethinking our patterns of thought but also considering that we're asking the wrong questions—thinking about the wrong things, the wrong patterns entirely—and we might be better served to approach a problem from an altogether different angle.

FITTING INTO AND BUILDING ON TECHNICAL COMMUNICATION RESEARCH

This project begins from a foundation of work begun by feminist technical communicators. In particular and as I detailed in chapter 1, the 1990s saw a series of special issues on feminisms by technical communication journals. These special journal issues provide the most systematic, discipline-sponsored engagements with feminism by technical communicators, and their relative absence since then—as though feminisms in technical communication were a completed project—is worrisome. However, my acknowledging this waning interest is not meant to discredit the important work of those relatively few scholars who have continued to publish at the intersection of feminisms and technical communication in more solitary, self-sponsored fashion. These scholars (Bellwoar 2012; Boyer and Webb 1992; Brady Aschauer 1999; Brasseur 1993, 2005; Coletta 1992; de Armas Ladd and Tangum 1992; Dragga 1993; Gregory 2012; Haas, Tulley, and Blair 2002; Hallenbeck 2012; Koerber 2002; Lay 1993; Lay, Monk, and Rosenfelt 2001; Lippincott 2003; Sauer 1993; Skinner 2012; Tebeaux 1993; Teston and Graham 2012) have kept many threads of feminist technical communication inquiry alive.

Feminist technical communicators have often discussed the need for more wide-ranging feminist approaches in our field. In particular, feminist technical communicators have:

- Persisted in doing important feminist work even in the absence of discipline- and journal-sponsored forums
- Embraced a plurality of feminisms and described myriad feminist methodologies and methods (Boyer and Webb 1992; de Armas Ladd and Tangum 1992; Haas and Eble 2018)
- Worked within and across gender studies, cultural studies, and social justice agendas (Brunner 1991; Carrell 1991; Flynn et al. 1991; Lay 1991, 1993)
- Conducted historiographic research and engaged in scholarly conversations about the importance of that research (Brasseur 1993,

2005; Durack 1997; E. A. Flynn 1997; J. Flynn 1997; Hallenbeck 2012; Lippincott 2003; Skinner 2012; Tebeaux 1993)
- Provided critical perspectives on technologies, sciences, terminologies, and social conventions that hide value systems in which misogyny is supported, valued, and reproduced (Bernhardt 1992; Bosley 1992; Brady Aschauer 1999; Colletta 1992; Dell 1992; Haas, Tulley, and Blair 2002; Lay 1993; Neeley 1992; Rifkind and Harper 1992; Sauer 1992; Tebeaux and Lay 1992)
- Engaged in interdisciplinary scholarship and activism (Bellwoar 2012; Gregory 2012; Koerber 2002; Teston and Graham 2012)
- Paid attention to the importance of social factors and intersectionality in terms of oppressions, theories, methodologies, practices, and more (Allen 1994; Bosley 1994; Dragga 1993; Gurak and Bayer 1994; LaDuc and Goldrick-Jones 1994; Ross 1994; Sauer 1994)
- Intervened in problematic actions (including rhetorics) that exist in and between public spheres, private lives, disciplinary venues, and pedagogical spaces (Lay, Monk, and Rosenfelt 2001; Sauer 1993).

Technical communicators should listen to and emulate this important work, and they also should be aware—and troubled—that this work represents a minority population and a diminishing disciplinary trend. Apparent feminism can help remedy this.

In addition, a significant body of work in technical communication and rhetoric that is not explicitly based in feminisms or gender studies or that doesn't make these theoretical approaches a main tenet is useful to the development of apparent feminisms—and I hope the reverse is true as well. For example, social justice technical communication work is attentive to feminist values but takes a broader approach (Jones 2016a, 2016b; Petersen and Walton 2018; Shelton 2019a, 2019b; Walton Moore, and Jones 2019). Further, the adjacent or umbrella (depending on your perspective) field of rhetoric offers a rich history of feminist work that not just influences apparent feminisms but also exists in conversation with, support of, and productive tension with feminist technical communication work (Ballif 2000; Bernard 1999; Enoch 2019; Flynn, Sotirin, and Brady 2012; Foss, Foss, and Griffin 1999; Glenn 1997; Hallenbeck 2012; Haraway 1996; Jarratt 2002; Logan 1999; Lunsford 1995; Ratcliffe 1995; Royster and Kirsch 2012). While a full review of feminist rhetorical literature is beyond the scope of this text, it is notable that technical communicators have often taken up the work of feminist rhetoricians and persistently share space with notable feminist rhetorical scholars at professional gatherings. In other words, the cross-pollination between feminist work in rhetorical studies and technical communication is rich.

Of particular importance to my thinking on apparent feminism is Krista Ratcliffe's (2018) exploration of narrative as a methodology in her afterword to Kristine L. Blair and Lee Nickoson's *Composing Feminist Interventions*, which suggests that the functions of narrative—to "explain to ourselves and to others what events we are narrating," "explain to ourselves and to others what we have learned about these narrated events," and "explain to ourselves and to others how we are constructing our own subjectivities (as points of view), the subjectivities of others (as characters in our own narratives), and the cultural spaces that we all share (as settings)" (506)—are complicated by the incorporation or acknowledgment of a feminist perspective. However, she then turns the problem on its head, suggesting that feminisms have helped scholars question the makeup of their disciplines and the narratives we tell about those disciplines, "identifying possibilities, limitations, and complications of narrative as a method of generating knowledge" (506). In thinking about what we make apparent, who we involve, and how we think of god terms like efficiency, it is important to keep at the forefront of our minds the sort of epistemological reflection Ratcliffe advocates.

APPARENT FEMINISMS AS/IN INTERDISCIPLINARY METHODOLOGY

Technical communicators have no choice but to be public intellectuals; knowing that our work will influence the public, then, we must go about it especially carefully. Melody Bowdon (2004, 325–326) recognized four major areas of responsibility, suggesting that as technical communicators we should (a) make our work part of the public sphere, (b) recognize the public nature of our work as teachers, (c) work toward positive change by recognizing social justice exigencies, and (d) recognize the subjectivity of our own bodies and being. Apparent feminism addresses the public sphere and attends to social, ethical, and political issues; works toward positive change in the world; and recognizes the existence of embodiment and its effects on theory. Bowdon also encouraged technical communicators "to tell the stories of these projects, to let our colleagues and students know about what's going on in these sites, and to identify places that might be good targets for more work" (339). As Claire Hemmings (2011) suggested, we do not even need to tell new stories; rather, we need to tell familiar stories differently. Apparent feminists can take up this call to tell such stories and thus insert feminist apparency into the public sphere. The first step in doing so is to pay attention to activists operating in the public sphere and to scholar-activists in complementary fields. In other words, apparent feminisms have important work to do in technical communication, but they grow from a fused foundation of both technical communication scholarship and work in adjacent fields.

Apparent feminism builds on and speaks with the technical communication scholarship surveyed above and in chapter 1. Further, I draw directly and heavily on five scholar-activists from outside the field—Indigenous environmental activist Winona LaDuke, cultural theorist and activist bell hooks (also known as Gloria Watkins), anthropologist Francesca Bray, queer theorist Jack Halberstam,[9] and transnational feminist theorist Chandra Talpade Mohanty—to assemble a methodology intended to help technical communicators attend to broad-ranging and far-reaching social, ethical, political, and practical needs. These scholars not only provide guidance for assembling an apparent feminist methodology; they also demonstrate how to employ it.

LaDuke's work spans several areas requiring further attention from technical communicators, including environmental research, ethnicity, and health rhetorics—all of which she connected to issues of gender. LaDuke provided a venue for valuing the voices and knowledges of Indigenous communities. Her work with communities whose health and lifeways have been threatened by legislative policy related to superfund sites is especially relevant. In her book *All Our Relations,* LaDuke (1999) chronicles the Mothers' Milk Project run by Mohawk environmental justice activist Katsi Cook, who "studied 50 new mothers over several years and documented a 200 percent greater concentration of PCBs in the breast milk of those mothers who ate fish from the St. Lawrence River as opposed to the general population" (19) as a result of General Motors (GM) using the area as a toxic waste dump. LaDuke shared Cook's efforts to heal the polluted land, water, associated food sources, and thus bodies of Mohawk people; she also traced multiple genres of technical communication—from meeting with and presenting to the Environmental Protection Agency (EPA) to facing down big industries, from conducting original scientific research to writing reports, from engaging with mainstream media to work focused on community healing—that could make a difference to them. Such work is a model to help move technical communicators toward being accountable for interrogating and responding to the potential injustices encoded in government and corporate policies.

Academic activist hooks's work is important for technical communicators to engage in theorizing visual rhetoric in relation to gender and ethnic studies. Although she is no stranger to rhetoric and composition, hooks self-identifies as a cultural critic and scholar of feminism, race, and social justice. Showing how the art world—a place where expression and entrance might seem more open—is dominated by white males and white male interests, hooks (1995) then turned to the creative work of Black artists such as photographer Carrie Mae Weems and painter Jean-Michel Basquiat to show how visual culture can be used for empowerment by addressing identity politics through art. By looking to these artists' past work for many of her examples of liberatory practices in visual culture, hooks both engaged in apparent feminist work and highlighted the importance of historiographic efforts.

Anthropologist Bray (1997) pointed out how technical communication is used to construct lives, spaces, times, and disciplines in gendered and power-laden ways. She took up historiographic goals in her feminist history of Imperial China; in so doing, she demonstrated that historiographic theory, methodology, and practice provide another way to support marginalized or underrepresented populations. Bray recovered and re-envisioned women's roles in domestic spaces, textile production, and reproductive technologies. Before doing so, she provided a framework for thinking about feminist historiography, articulating a systematic methodology for undertaking gender-based inquiry. Bray theorized gynotechnics, a methodology that involves recognizing "a technical system that produces ideas about women, and therefore about a gender system and about hierarchical relations in general" (4).

Queer theorist Halberstam's (2005) reimagining of time and space has profound implications for how we understand the production and distribution of technical documents; this queer reimagining is necessary to make the limitations of gender more apparent. An example of technical communication constructing our lives is the signage for public bathrooms, which is almost exclusively produced based on the assumption of sexual dimorphism. Thus, non-normatively sexed and gendered people are confronted with an uneasy choice when faced with public bathrooms and other public spaces that are constructed based on sex or gender. Halberstam further argued that technical documents such as marriage contracts reify normative perspectives about the time line for and required events in a person's life. As technical documents restrict our understandings of how, where, and when we do things, this area of inquiry is another space for interrogating the purposes, goals, biases, and implications of technical communication as seen through the lenses of gender studies and feminisms.

Mohanty's work introduces productive problems for feminisms. In her famous essay "Under Western Eyes: Feminist Scholarship and Colonial Discourses," Mohanty (1988), writing from a place of "implication and investment," critiques the notion of the monolithic "third world woman" in the imaginations of many Western feminists. She argued that the construction of a monolithic Other is a colonizing act and notes that the same critique applies to the category of "woman"—an issue also of great significance to queer and trans communities:

> The relationship between "Woman"—a cultural and ideological composite Other constructed through diverse representational discourses (scientific, literary, juridical, linguistic, cinematic, etc.)—and "women"—real, material subjects of their collective histories—is one of the central questions the practice of feminist scholarship seeks to address. This connection between women as historical subjects and the re-presentation of Woman produced by hegemonic discourses is not a relation of direct identity, or a relation of correspondence or simple implication. It is an arbitrary relation set up by particular cultures. I would like to suggest that the feminist writings I analyze here discursively colonize the material and historical heterogeneities of the lives of women in the third world. (62)

Put another way, Mohanty pointed out that feminist studies is interested in the material lives of women but that it has not always done critical work in defining its subjects. Exactly whose materials lives are we talking about? Who are women? Who counts? What assumptions do we make when the word *woman* or the term *third world women* are invoked? Importantly, Mohanty's objection to Western feminists' reductionism in their use of the term *third world women* does not foreclose the possibility of coalition, which she says directly in a 2003 retrospective on the original article. Rather, Mohanty argued that the only way out of this bind is through coalition, diversity, and constant struggle. Tellingly, this means a turn to people rather than to theory (if the two can in fact be separated). In "Under Western Eyes Revisited," Mohanty (2003, 530) spends a considerable amount of time on pedagogy (a departure from the original essay), arguing that "feminist activist teachers must struggle with themselves and each other to open the world with all its complexity to their students."

Mohanty, Halberstam, Bray, hooks, and LaDuke point us toward a conversation about social constructionism, cultural relativity, and social justice. This is a complicated debate, one whose largest value may lie in its persistent continuation. Apparent feminism, as a methodology, does not stake a single firm position in this debate. It is proactively interested in social justice, sympathetic to social constructionism, and attentive to cultural difference. Although I find most approaches to posthumanism troubling because of the privilege necessary to make this turn when so many humans continue to be deeply oppressed, Diana Coole and Samantha Frost's (2010, 27) explanation of posthumanism offers an instructive approach:

It is entirely possible, then, to accept social constructionist arguments while also insisting that the material realm is irreducible to culture or discourse and that cultural artifacts are not arbitrary vis-à-vis nature . . . it does not follow that . . . acknowledging nondiscursive material efficacy is equivalent to espousing a metaphysical claim regarding the Real as ultimate truth. For critical materialists, society is simultaneously materially real and socially constructed: our material lives are always culturally mediated, but they are not only cultural. As in new materialist ontologies, the challenge here is to give materiality its due while recognizing its plural dimensions and its complex, contingent modes of appearing.

Charting a middle path, it seems, is not only possible but necessary. The difficulty of navigating a position that is not theoretically pure, however, requires constant vigilance—and the willingness to let go of that theoretical orientation when it outlives its usefulness.[10]

The five scholar-activists whose work I discuss above (LaDuke, hooks, Bray, Halberstam, and Mohanty) are the most important influences I drew on in developing apparent feminisms. However, a number of other scholars from disciplines beyond technical communication and rhetoric also provide pathways for understanding the important opportunities for dialogue opened by apparent feminisms. These conversations can impact the ongoing formation and development of this theory. For example, in their attention to the risks one runs with apparency, apparent feminisms ask us to pay attention to social roles—not necessarily for the purposes of critique but for the purposes of awareness.

As an example of attention to social roles and their importance to apparent feminist thinking, in her short commentary on the social effects and causes of ectogenesis (the growing of, in this case, human embryos in vitro), Lisa Campo-Engelstein (2020, 85) argues that in vitro fertilization (IVF) is an example of how "technomedical approaches generally do not solve social problems because they do not address the root of the issue, which is social in nature, not medical." While IVF may increase the autonomy of individual, privileged women, it does nothing to alter or subvert the social norms surrounding parenting roles. This does not mean that women with access to IVF should not use it. It does mean that women with access to IVF have an obligation to be aware of their relative social privilege.

Anthropologist Emily Martin (2001) can help us take a step even further beyond recognition of social roles: she discusses the very American nature of *having a body*. According to Martin, "[Kmer refugee-immigrants to the US] see acquiring a biological body as a specifically American experience: to act as an American, and to become a citizen, one must become an inhabiter of this thing called a body" (xxvi).

This tells us that the relationship of US women to their bodies is not a default and can shed light on Western women's "tortured efforts to reconcile experience with medical expectations" (89). In a parallel to the ways social systems obviate the presence of the laborer, Martin finds that some medical texts from the twentieth century largely ignore the existence of women when explaining labor (147). Martin also discusses the correlative effect of race and socioeconomic class on both maternal and fetal mortality (148): "The causes of these differentials surely lie within the different social circumstances of different groups" (149). In short, Martin finds that women's social roles provide them with a unique and important perspective, which supports my contention that feminisms—as one option among a variety of biased worldviews—are a particularly useful perspective. Martin says women are engaged in housekeeping—of their own bodies and of their homes—in ways men are not, and this makes them "more able to see the social whole, which includes both its concrete and its abstract parts" (201).

Sara Ahmed's *Queer Phenomenology* (2006) is a work emerging from feminisms, queer theory, and postcolonialism that can help guide apparent feminists in thinking about the orientations particular definitions of efficiency preclude. To explain, Ahmed asks what it means "for sexuality to be lived as orientated" (1) and argues that using orientation as a lens can help us to understand and theorize how spaces, places, and time (line)s are implicitly sexualized. Using the table as an extended example, Ahmed shows how our orientations toward an object can build our ideas about our lives. She also invokes space as an important concept in studying orientation: "Space acquires 'direction' through how bodies inhabit it, just as bodies acquire direction in this inhabitance" (12). We might theorize our orientations toward a body of water, a medical procedure, or a public health system. Ahmed points out that "when things are orientated they are facing the right way: in other words, the objects around the body allow the body itself to be extended. When things are orientated, we are occupied and busy" (51). However, "What happens when we are 'knocked off course' depends on the psychic and social resources 'behind' us" (19). This, of course, means that people inhabiting bodies with fewer resources have fewer options and may be more easily or permanently disoriented. What constitutes an event that might knock someone off course is open to the imagination; such events extend into any sphere of life and do not respect the boundaries of those spheres: "Acts of domestication are not private; they involve the shaping of collective bodies" (117). Feminisms, particularly when rendered purposefully apparent and fallible, can be iteratively attentive to such difference.

Differences—constraints—can be a defining factor in world making. This does not necessarily need to be a negative thing, however. N. Katharine Hayles (2019, 37) points out that "life depends on the continuing existence of many different kinds of constraints, from cell membranes to energy and temperature requirements." Championing biosemiotics, Hayles argues that we must learn that meaning making is not a human-exclusive endeavor. Meaning making can come from sources we may not have considered; the fact "that biological brains use input from their environments, sensory systems, and bodily functions to achieve cognition does not mean that computers must also be embodied and enworlded in the same way as humans to engage in meaning-making practices" (42–43). Hayles argues for new perspectives with the potential to identify and acknowledge contributions and theoretical frameworks that emerge from processes that are unfamiliar. She says that "computational media act as ethical agents in our contemporary world" (53). This echoes my discussion of posthumanism above, and I admit to the same reservations and excitement about this line of thinking, with one significant addition: Hayles specifically looks to "theoretical frameworks that underscore the importance of cognitive media in creating the meanings that guide hybrid human-technical action, perception, and decision-making in the contemporary world-horizon of the cognisphere" (51–52). This specific invocation of the technical can help apparent feminists make and hold space for ways of being beyond our ability to imagine and also to more quickly apply those ways of being in the oft-overlooked technical spheres that are such important sites of culture building.

Nayantara Sheoran Appleton (2018) provides an important complement to Hayles's work. Drawing on Donna J. Haraway (1991) and Emily Martin (1991), Appleton warns against uncritical politics of inclusion. While inclusion can be an important goal, it can also fool us into forgetting to be troubled, critical, careful. Appleton differentiates two feminist responses to masculinist traditions in science: on the one hand, having more women at the table and, on the other, having women at the table who change the substance of science. She gives examples of when women "have changed the substance of science, as opposed to just participating more, dare I say, 'leaning into' a patriarchal science" (145). While it's hard to imagine that having women at the table does not itself change the "substance of science," Appleton's point that a radical politics of inclusion looks different from increased participation is well taken:

> The feminist commons for techno-science is one place to connect with science and scientific knowledges, but it is a space that is a starting point as opposed to an end point. It opens the doors for engagements with

queer, indigenous, and marginalized communities as sites for learning and contributing to, rather than spaces to colonize for raw material for knowledge production. In offering to always be situated *and* partial, the knowledge that emerges from such scientific commons is feminist and not attached to femininity or the limited male/female binary opposition. To be situated in the feminist commons *is not to be stuck*, but rather [to] have a grounding that allows for critical engagements with scientific knowledge and its production. (147, original emphases)

Another way of thinking about this is to imagine the earlier dichotomy—being at the table versus changing the substance of the conversation—as existing on a temporal continuum; it is sometimes important or even necessary to be *included* before one can engage in cultural *change*. Apparent feminisms do not take a particular side in this debate but rather encourage a rhetorically informed analysis of which position (or combination thereof) might be most effective in a given situation at a particular point in time.

Finally, I am not the only scholar to argue for the importance of rhetorical approaches to feminisms. In a recent example of such work, Cassandra Woody (2020) uses the term *procedural feminism* to advocate for something very much like an apparent feminist approach to first-year composition (FYC), noting that having students enact rather than study feminism circumvents their knee-jerk reactions to the label—a label that, to them, is "an extension of elite liberalism that butts up against their conservative values" (483). For Woody, a pedagogical turn is appropriate, and she notes that programmatic planning can benefit from an approach where "feminist theory that informs classroom activity must be embedded within curricula rather than the subject matter students engage" (483). While Woody seems to somewhat conflate feminist practices and women's practices, she also rightly points to the importance of both in getting students to think outside and beyond themselves. In particular, her attention to Ratcliffe's rhetorical listening and her frank assessment of standard FYC practices offer useful inroads for apparent feminist pedagogues in a variety of contexts: "When writing courses rooted in rhetorical education keep students focused on public issues and require research to involve peer-reviewed sources or sources easily found in public circulation, the classes tend to maintain the hierarchy that privileges male-dominated domains. Using feminist rhetorical practices that encourage slow argument and emphasize overlooked spaces within which rhetorical activities form to guide writing curricula, however, can 'disrupt the "seamless narrative"' (Lunsford 1995, 6) that devalues counterpublic arenas and creates false binaries" (489). Ultimately, Woody calls for a feminism that reimagines rhetoric's purpose as dialogic but not necessarily persuasive. This differs from apparent feminisms, which embrace persuasion—just not at the expense of dialogue.

REASSEMBLING A THEORY OF FEMINIST APPARENCY

While I have articulated this theory of apparent feminism as a sort of three-pronged approach, it is important to understand that these three goals are interrelated, and each is reliant on the others. In the following list, I demonstrate these interrelationships by specifically articulating one way each goal relies on the other two.

- The goal of making more apparent to as many audiences as possible the need for feminist interventions depends on
  * having allies to help us widen our reach, and
  * understanding apparent feminism as an efficient approach.
- The goal of hailing non-feminists as allies depends on
  * making our feminism apparent so that allies might recognize us, and
  * believing that attracting allies is an efficient approach.
- The goal of demystifying the relationship between feminism and efficiency depends on
  * making apparent our reimagining of the term *efficiency* and of technical communication as a discipline, and
  * having allies who can participate in this reimagining so it has effects in social, ethical, political, and practical domains.

This list makes clear how each goal of this methodology relies on the other two goals. Still, this is not an exhaustive list of the ways these features interrelate but rather a sketch intended to show some of the ways each goal depends on the others.

I call on technical communicators to take up this methodology so they can point out and intervene in systemic oppression. The examples I have given here—most prominently the Texas Woman's Right to Know Act but also similar laws passed by Arizona and Virginia—are just a small sample of exigencies that bear specifically on reproductive justice (Arizona Women's Health and Safety Act 2012; SB 484 2012; Texas Health and Safety Code 2012).[11] Apparent feminism can be leveraged to address a wide range of social injustices in a wide range of arenas all bearing on technical communication, from government regulations to social policy to digital rhetorics to professional gatekeeping to street harassment and more.

Apparent feminism is an urgently needed methodology in technical communication at this kairotic moment precisely because of the knee-jerk response that makes feminism unpersuasive to some audiences; conversations about feminism, particularly in technical realms, too often result in it being vilified or dismissed as biased. It is more important than ever for feminist work to be marked as such in the face of allegations of a post-feminist world, particularly in male-dominated disciplines such as technical communication. Feminist technical communicators "must continue to seek out ways of thinking that will help make operations of power and hegemony" more apparent (LaDuc and Goldrick-Jones 1994, 247). Ultimately, I hope apparent feminism can provide a temporary, permeable, and flexible approach to the many current situations facing technical communicators that are in need of feminist intervention.

3
SLOW CRISIS

Having established the basic shape—although it is a temporary and permeable one—of apparent feminisms, I now turn to a more concerted focus on the tenet having to do with efficiency. This tenet leads us directly to an understanding of a phenomenon I call *slow crisis*. This chapter explains how apparent feminist critiques of efficiency beget slow crisis. In the remainder of this book, I utilize the concept of *slow crisis* as it relates to the Deepwater Horizon Disaster to show one potential apparent feminist view of efficiency. I aim to demonstrate how a different understanding of efficiency makes apparent the reality of slow crisis and how this could result in better technical communication and more just and equitable technical practice.

APPARENT FEMINIST EFFICIENCY AND SLOW CRISIS

To explain how apparent feminist critiques of efficiency invest slow crisis with meaning in this context, I take a rhetorical view of risk and invest in eco-critical and ecological-economic perspectives. Risk communication—and rhetorical perspectives on it—could be the subject of an entire book itself (and has been), so here I offer a very brief review of risk communication in the field of technical communication to help ground the terministic screen (slow crisis) I discuss in this chapter and model in the next chapter.

Over time, technical communicators have worked to showcase and better examine the importance of social and cultural factors in the evaluation of risk. Jeffrey T. Grabill and W. Michele Simmons (1998) argued for a critical rhetoric of risk communication, suggesting that risk is socially constructed. They suggested that definitional disputes in risk communication situations "are a public contestation over the meaning of risk—the 'truth' about risk is actually a product of such disputes" (423). Grabill and Simmons advocated for citizen participation in the construction of rhetorics of risk. They were, in part, taking up a call made by Alonzo Plough and Sheldon Krimsky (1990), who expressed the need for situating risk communication within a cultural framework. Plough and Krimsky asserted that "if the technosphere begins to appreciate and respect the logic of local culture toward risk events and if local culture has access to a demystified science, points of intersection will be possible" (229–230). As shown by the works of Plough and Krimsky and of Grabill and Simmons, technical communicators must consider the implications of believing that risk exists independent of culture. If we do not acknowledge the cultural power communicators exert in constructing understandings of risk, then we may well fail to see the ideologies driving those kinds of persuasive, risk-constructing communications. This is, in effect, an apparent feminist view of risk; risk is culturally situated and saturated and, as we will soon see, it calls for feminist reimaginations.

One way to make visible the mechanizations of power is to return to the idea of location. According to Michel de Certeau (1984), strategies emanate from a rooted base of power. Building on the work of de Certeau, Huiling Ding (2009, 330) theorized "guerrilla media," which are mobile or digital technologies functioning "as 'weapons of the weak' that open up media spaces to marginalized groups and help define new political situations." The rhetorical function of guerrilla media is to take over "readily accessible" cultural sites that can "serve as one possible entry point into power systems for tactical intervention to challenge or contradict dominant discourses" (344).

I extend this argument by suggesting that an important kind of work guerrilla media always engages in is the construction of risk. This construction includes assumptions about the framing of time and its relative importance in any given hazard situation. The existence of guerrilla media points to the attempted silencing of some perspectives within dominant discourses and, in so doing, communicates the risk of being silenced—especially at particular junctures—on an issue that matters. Sonia H. Stephens and Daniel P. Richards (2020) offer a useful review of risk literature. They also delve into the complex question of how to quantify risk: "Where risk has traditionally been defined by experts as the probability of a hazard's occurrence times the size of its impact . . . contemporary research shows that public perception of risk is multidimensional . . . and includes a risk's impact, an individual's confidence in scientific understanding of the risk, and perceived dread" (5). Peter M. Sandman (2020) has argued that risk equals hazard plus outrage. Daniel R. Wildcat (2009, 127) advances a similar equation—"$3C/E=T$, also expressed as community times communication times culture divided by environment equals technology"—as a way of thinking about risk situations. One factor that is difficult to quantify in any of these conceptualizations is time. At what point should we run a given formula? Is each of the above formulae designed to function at the point of crisis, or after it, or before it? What is that point, and who decides?

One theoretization that does explicitly mention time is an apparent decolonial feminist approach to risk (Haas and Frost 2017). This approach builds on the importance of social and cultural considerations highlighted in theories above and adds a component that "makes connections between historical, contemporary, and future risks" (169). Indeed, interfacing an apparent decolonial feminist approach with questions about efficiency helps us think critically about time and points to the ways our approaches to time are the result of hierarchical thinking. Grabill and Simmons's (1998) imperatives to reframe risk as produced by a complex network of participants, to make more visible the processes and locations of decision making in risk communication, and to contextualize and localize risk communication practices can help researchers pay attention to hierarchies—including hierarchies that privilege particular perceptions of time—that preexist the observable situation. For example, local bloggers can use rhetorical means to construct risk in globalized spaces, but larger collective entities are still privileged by internet search engines. According to Liza Potts (2009, 295), "Search algorithms are based on rewarding established content. Disaster content is brand new"—although disaster content on an already established site like BP's has an advantage in terms of coming up in searches. In addition to the culturally loaded results of online searches, we can see layered differences in actants within networks by considering what Arjun Appadurai (2000, 15) called "grassroots globalization" wherein those operating from "below" were "concerned with mobilizing highly specific local, national, and regional groups on matters of equity, access, justice, and redistribution." Local actants who put information online did so because they had culturally significant meaning making to discuss that was not being articulated elsewhere. Apparent feminist approaches to efficiency provide a framework for interrogating how actors in this system determine whose constructions of risk are taken up and disseminated. This enculturated approach can also focus on temporal difference and its importance in evaluating risk.

In arriving at a context for application—the Deepwater Horizon Disaster—it would be impossible to do an informed analysis without acknowledging eco-critical and ecological-economic perspectives. Ecocriticism's main project is environmental awareness, and it is often associated with modern trends toward sustainability activism. Ecological economics is concerned with the complex relationship between human economies and eco-critical issues. It "seeks to ground economic thinking in the dual realities and constraints of our biophysical and moral environments" (Daly and Farley 2004, xxi). The construction of both ecologic and economic risk associated with the Deepwater Horizon Disaster was and is very much mediated within and across transcultural networks; indeed, the tension between these two different measures of risk is where much of the conflict between local and non-local rhetorics lies. The importance of listening to and thinking through such conflicts is proven by past activist work that has enabled marginalized voices and cultures to be heard transculturally on ecological issues—which returns us to apparent feminisms' transdisciplinary relations. For example, Winona LaDuke (1999) used an eco-critical lens in her examination of the environmental effects of PCBs, superfund sites, nuclear waste, and other contaminants on Native American lands. She questioned, "Who gets to determine the destiny of the land, and of the people who live on it" (5). The combination of LaDuke's arguments with Daly and Farley's, as well as a rhetorical perspective of risk, shows that analyzing economic agency without recognizing its attached ecological and ethical effects over time can be extremely dangerous. This recognition of the interconnectedness of ecology, economy, rhetoric, and ethics demonstrates that in examining what is articulated as efficient, we may need to expand our temporal awareness to ask better questions. Apparent feminist efficiency leads to and is opened up by the notion of slow crisis.

SLOW CRISIS

If we are to apply feminisms to the seemingly objective realm of technical communication, it makes sense to ground this feminist urgency in something tangible. That is, feminist quarrels with the false objectivity of technical communication are most persuasive when grounded in lived experience. It also makes sense that feminist approaches to technical communication will support and be bolstered by queer approaches, which also value non-normative epistemologies. In fact, modern feminisms that aim to be inclusive and progressive should always already be working in concert with a variety of progressive approaches. Queer approaches, in particular, can offer new and alternative ways of seeing and structuring lived experience, and they run complementary to—often overlapping—feminist perspectives. In what follows, I delve into lived experience—indeed, visceral experience—and utilize both feminist and queer approaches to restructure it.

To better explain how and why we should ground feminist urgency toward technical communication in lived approaches, I offer a story—one that is difficult to bear witness to and all the more urgent for that reality. Peggy Stewart married Mike Stewart in 1974. He viciously abused her—physically, sexually, mentally. He put her in the hospital on at least one occasion, he shot her cat, he abused her daughters and tried to force Peggy to kill and bury one of them, he threatened her life with a gun to her head. Peggy sought in-patient psychiatric treatment; Mike removed her from the hospital and forced her to return home. To put it succinctly and mildly, Peggy was a victim of domestic abuse. The day after Mike forced Peggy home from the hospital, Peggy found a loaded gun in the house and, afraid of what Mike would do with it, hid it under a mattress. In the midst of an abusive episode, Mike told Peggy not to bother cleaning the house because she wouldn't be there much longer. When he went to sleep, Peggy, fearing he would kill her when he awoke, retrieved the gun and killed him while he slept (Krienert, personal communication, 2011; Ogle and Jacobs 2002; *State v. Stewart* 1988).

Peggy's counsel prepared a self-defense claim. The standard for self-defense varies from state to state, but often the requirements hinge on (1) reasonableness and (2) imminence. To explain, a juror might have to decide if a reasonable person would believe Mike was going to murder Peggy—or, alternately—depending on the interpretation of "reasonable"—a juror might have to decide if Peggy, given her circumstances, reasonably believed Mike was going to murder her. This contextual distinction is, of course, very important—particularly with the use of Battered Woman Syndrome in some legal defenses. Battered Woman Syndrome, first described by psychotherapist Lenore Walker in the 1970s, explains the posttraumatic-stress–like symptoms often experienced by victims of domestic abuse and can be used to explain behaviors that otherwise appear nonsensical to those not in the relationship. Some criminal justice scholars take exception to Battered Woman Syndrome, arguing that the battered people in question are more qualified than any other person to read their abuser's cues and infer the actions about to occur. In other words, arguments could be made for abused partners having either compromised judgment or expert judgment.

In terms of the second criterion, some courts define imminence to mean that a confrontation must be occurring at the moment of the homicidal act. Others disagree, and some question what constitutes a "moment." Still others have held that the terms *immediate* and *imminent* mean different things, with immediacy sometimes used as the standard and interpreted more broadly to mean that a confrontation was occurring over a period of time.

Both of these standards appear to us to be cloaked in the professionalism, the unimpeachability, of legal rhetoric. Legal rhetoric, as we know, can be impenetrable to the average person and also has a history of protecting the interests of those who developed it (Williams 1992). What is legal is not always ethical, moral, equitable, or fair. Critical race theory, in particular, can demonstrate a long history of critiques aimed at legal rhetorics that were inherently discriminatory and overtly racist, sexist, or both; the very concept of intersectionality emerged from and in response to legal rhetorics (Crenshaw 1991). These arguments, though seemingly highly intellectualized, are also visceral—more so for some than others. The ways these standards are (or are not) applied determine the fates of women like Peggy Stewart.

Peggy's case is not unique. A number of cases have taken up the question of how to judge a woman who shoots her abusive partner in a moment of vulnerability (see, just to begin, *People v. Aris* 1989; *People v. Beasley* 1993; *State v. Norman* 1989; *State v. Stewart* 1988). However, there does not appear to be a firm legal precedent on the matter, and that is in part down to the question of the second standard described above: imminence, immediacy, or both. Different courts interpret this standard and these terms in different ways.

This standard of immediacy is the one that is of concern to us here, and the variance in interpretation is where I ground an apparent feminist intervention. For whom is the imminence standard, under its strictest interpretation, efficient? Well, for abusers. If we interpret imminence to mean that a battered wife can only shoot her husband in the moment before he pulls the trigger himself, we put him at a supreme advantage in terms of both survival and avoiding punishment. This interpretation is efficient for an abuser because it adds one more barrier to the already nearly insurmountable set of barriers preventing an abused partner from defending themselves.

A broader interpretation of immediacy shifts the answer to the apparent feminist question, "for whom is this system efficient?" Indeed, a broader interpretation ultimately invokes the concept of "slow homicide." This argument suggests that Mike Stewart was indeed engaged in the act of murdering Peggy Stewart. We can understand "the battering relationship as a homicidal process" (Ogle and Jacobs 2002, 70); Mike Stewart was engaged in committing a slow homicide. The murder was happening over an extended period of time, but it was inevitable. If, for example, we are to accept self-defense in relation to a mugging that took thirty seconds, a robbery that took three minutes, and also a kidnapping that took place over thirty minutes when the victim finally freed herself and reached a weapon, then at what point is immediacy no longer applicable? Mike Stewart may not have been actively strangling, drowning, or beating Peggy Stewart at the time she shot him, but she knew that he would kill her and she chose the only moment when he did not have a size and strength advantage to defend herself. She defended herself from Mike Stewart's act of slow homicide.

In essence, slow homicide suggests that cultural context must be taken into account. Robbin S. Ogle and Susan Jacobs (2002, 70, 78, original emphasis) argue that "cultural, social, and structural issues" are integral to understanding battering relationships and the ways they are maintained and repeated (70) and that the "contextual or *social framework* is essential to understanding the battering relationship as an ongoing homicidal interaction." Ogle and Jacobs distinguish between common couple violence, which is limited to stressful situations, and patriarchal terrorism, which is when one partner works to exert complete control over the other. This social interaction perspective relies on the fact that the victim lives in constant fear. Victims, like Peggy, are always—correctly—attuned to the possibility of violence.

At the heart of this conflict is a difference in understandings of time. The courts have begun to question their past temporal bias toward understanding imminence in a narrowly conceived framework. In a similar fashion as these legal defenses, we might argue that the parameters of the term *crisis*—as applied in a wide variety of contexts, not just battering relationships—depend largely on a person's perspective. The moments that bound the beginning and ending of a crisis might not be as clear as we would like to believe. "Slow" patterns actually constitute crisis under some conditions—conditions that might include, for example, not only spouse battering but also systematic assaults on reproductive rights, human health effects of slowly unfolding environmental disasters, and shifts in what is considered "standard practice" over time. In fact, slow crises are already widely recognized in some spheres; Bess McCullouch (personal communication, 2021) notes that masculine-sphere spaces such as military intelligence recognize the importance of, say, long-term monitoring of a hostile country that has potential nuclear capability without any quibbling.

While precedent in the legal system is hardly at the cutting edge of academic reconsiderations of temporality, the legal theorization of slow homicide is instructive. In the same ways slow homicide alters our temporal perspective to help us see something that was always present but not immediately apparent, we can think about *slow crisis* as an event that produces exigencies but that unfolds on a temporal basis we might not take as default. In other words, crises can manifest slowly, just as homicides can.

To better articulate what the concept of slow crisis affords us as technical communicators, I turn to a similar concept introduced by Rob Nixon (2011b): slow violence. Nixon theorizes slow violence along some of the same axes I identify as values for those who might take up slow crisis. He points to environmental concerns specifically. "By slow violence I mean a violence that occurs gradually and out of sight, a violence of delayed destruction that is dispersed across time and space, an attritional violence that is typically not viewed as violence at all" (2). Nixon also identifies slow violence as not just attritional but exponential, calling forth ideas of slow change over time going unnoticed even as the *rate* of change also goes unnoticed: "Attritional catastrophes that overspill clear boundaries in time and space are marked above all by displacements—temporal, geographical, rhetorical, and technological displacements that simplify violence and underestimate, in advance and in retrospect, the human and environmental costs. Such displacements smooth the way for amnesia, as places are rendered irretrievable to those who once inhabited them, places that ordinarily pass unmourned in the corporate media" (7). Nixon argues that environmental violence "needs to be seen—and deeply considered—as a contest not only over space, or bodies, or labor, or resources, but also over time" (8). Famed environmental activist Rachel Carson called a similar concept "death by indirection." The contexts in which death of indirection or slow violence are enacted are ubiquitous but also include some common topics. Indeed, Nixon directly addresses the context this book takes on—the Deepwater Horizon Disaster—by pointing to time as a critical point of contestation, with some actors invoking the notion of recovery over time as a defensive mechanism: "Big Oil and government agencies both invoked natural resilience as an advance strategy for minimizing oversight" (21). In this way, the passage of time is made out as healing and human influence is minimized—a position that is contradicted by the theorization of slow homicide, which argues that an actor retains agency even when/if their actions take place over an extended period of time relative to human perception of "crisis."

This focus on the importance of time is not novel among either philosophers or environmental activists. Both Jean Baudrillard and Lauren Berlant theorize "slow death," with Baudrillard's work based in economics where slow death is used to refer to the double bind of the capitalist subject who is bound to both labor and consumerism. For Baudrillard (1976, 167), use-value means utility and exchange value means market price:

> Value, in particular time as value, is accumulated in the phantasm of death deferred, pending the term of a linear infinity of value. Even those who no longer believe in a personal eternity believe in the infinity of time as they do in a species-capital of double-compound interest. The infinity of capital passes into the infinity of time, the eternity of a productive system no longer familiar with the reversibility of gift-exchange, but instead with the irreversibility of quantitative growth. The accumulation of time imposes the idea of progress, as the accumulation of science imposes the idea of truth: in each case, what is accumulated is no longer symbolically exchanged, but becomes an objective dimension. Ultimately, the total objectivity of time, like total accumulation, is the total impossibility of symbolic exchange, that is, death. Hence the absolute impasse of political economy, which intends to eliminate death through accumulation: the time of accumulation is the time of death itself. We cannot hope for a dialectical revolution at the end of this process of spiraling hoarding.

Baudrillard's reflections might remind us of a recent meme that has circulated in protest to material accumulation, claiming that "all this *stuff* was once *time*."

Meanwhile, Berlant's (2011, 95) "slow death" is predicated on the notion of inevitability: "The phrase slow death refers to the physical wearing out of a population in a way that points to its deterioration as a defining condition of its experience and historical existence." This iteration of slow death might recall the situation of populations that live for generations in poverty, unable to break a cycle of government/social neglect and always already on the front lines of environmental disaster and risk (see Haas and Frost 2017). It might relate to the always already inferior position of women, who must work twice as hard for half as much, in the worlds of business and law. It might refer to the exhaustion felt by people of color in a nation that degrades and destroys their lives and bodies with impunity. Berlant (2011, 117) explains further:

> I am focusing here on the way the attrition of the subject of capital articulates survival with slow death. Impassivity and other politically depressed relations of alienation, coolness, detachment, or distraction, especially in subordinated populations, can be read as affective forms of engagement with the environments of slow death, much as the violence of battered women has had to be reunderstood as a kind of destruction toward survival . . . The structural position of the overwhelmed life intensifies this foreshortening of consciousness and fantasy; under a regime of crisis ordinariness, life feels truncated, more like desperate doggy paddling than like a magnificent swim out to the horizon.

In short, Berlant (like Baudrillard) frames slow death as a concept that helps us think about material existence at the same time as political structures. Berlant, though, makes an explicit connection to the important example—domestic abuse—from which I extrapolated slow crisis above. In so doing, she underscores the importance of understanding how the victims of what I would term crisis are always already oppressed.

Nixon says slow violence differs from structural violence in that it "has a wider descriptive range in calling attention, not simply to questions of agency, but to broader, more complex descriptive categories of violence enacted slowly over time" (2011b, 11). That is, he is expanding one of the axes along which structural violence is commonly recognized. Nixon (2011a) recognizes the applicability of a shifted temporal frame:

> We are accustomed to conceiving violence as immediate and explosive, erupting into instant, concentrated visibility. But we need to revisit our assumptions and consider the relative invisibility of slow violence. I mean a violence that is neither spectacular nor instantaneous but instead incremental, whose calamitous repercussions are postponed for years or decades or centuries. I want, then, to complicate conventional perceptions of violence as a highly visible act that is newsworthy because it is focused around an event, bounded by time, and aimed at a specific body or bodies. Emphasizing the temporal dispersion of slow violence can change the way we perceive and respond to a variety of social crises, like domestic abuse or post-traumatic stress, but it is particularly pertinent to the strategic challenges of environmental calamities.

Violence, then, is not always immediately apparent—and this lack of apparency begets cycles in which it becomes less and less obvious.

I differ from Nixon in that I find the term *slow crisis* to be more instructive and useful than slow violence. Women have long recognized the horror of slow violence; violence has always happened at many speeds for us. Crisis, though, suggests urgency and speed. Rhetorics of efficiency sponsor an urgency to identify and resolve crises—at least the ones that are recognized as such (efficient for whom?)—as quickly as possible. I suggest that technical communicators must do a better job of recognizing and resisting this rush to efficient resolution of obvious crises and instead recognize that slow patterns can also constitute crisis. We must not let the urgency of our attention to the former outweigh the importance of our attention to the latter.

More important, I also diverge from Nixon in that I vehemently resist the deconstruction of the connection between violence at whatever speed and the bodies such violence is inflicted upon. That is, it is important that we not take the anecdote that opens this section and the parallel I use it to introduce lightly. We must not pull the theoretical basis of Peggy Stewart's self-defense claim away from the horror of her lived reality. We cannot take an idea used to describe what has been done to women's bodies and animate that idea for non-human actors without careful thought. Connections between women's bodies and the environment are ubiquitous (Griffin 2016). However, women's bodies and the historical violence enacted upon them—including through environmental means—cannot and should not be separated for the sake of theory.

Thus, some critical differences between slow violence and slow crisis make the latter term necessary for the apparent feminist critique in this book. First, slow crisis is a term I employ only when it is always already paired with a feminist perspective; borrowing from the term *slow homicide* and the rich, terrible context in which that notion was formed otherwise feels ethically suspect. Further, whereas violence suggests an actor, an agent perpetuating some threat or injustice, crisis does a better job of capturing the possibility of passive threat—threat by refusal to actively change or resist the status quo. Articulating a crisis as acting slowly, over a temporal field that is not our usual perspective, does not require (although it can allow for) casting a villain. This is important for apparent feminist critiques because their target is so often the status quo—not rhetors themselves, who are often victims of their socialization, but the technical documentation so long presumed to be neutral, objective, transparent. This said, the notion of slow crisis does not preclude accountability—the laying of blame is sometimes necessary in instances where a critical agent actively refuses to resist the status quo.

QUEER TEMPORALITY

Theoretical explorations can sometimes have the perceived effect of leaving bodies behind. We become immersed in theory and imagine ourselves as minds that transcend the physical bodies they are housed in. Drawing on William P. Banks' (2003) contention that modernist discourse erases discourse-producing bodies, A. Abby Knoblauch and Marie E. Moeller (2022, 8, original emphases) state that "the body is often only seen *as a body* when it is not the presumed norm. Knowledge, then, is often only seen as *embodied* when the body producing that knowledge is imagined as Other." The notion that the mind can be separated from the body—the so-called Cartesian split—is indicative, of course, of a fallacious sense of self. Even setting aside the biology of the brain, our corporeality begets certain experiences that, in turn, shape our thinking: "A mind is never at work on its own: minds are always operating within and inseparably from full humans with complex interpersonal, bodily, and emotional intersectionalities" (Owens and Molloy 2022, 241). Theory without embodiment is not just incomplete; it is incorrect. In some contexts, this means that projects to return data to bodies are urgent (Frost and Eble 2020). In others, it means we must proactively seek theories that acknowledge, value, and listen to embodiment as a source of knowledge.

To fully theorize the notion of slow crisis while resisting any related drawing away from embodied experience, I turn to queer theory. Queer theorists have long recognized the violence done to a wide variety of people by social expectations to live our lives on a certain time line. For example, the expectation to marry is problematic for a person who is not allowed to marry, marriage equality having been secured[1] in the United States only in 2015 by the US Supreme Court in *Obergefell v. Hodges*. The expectation to reproduce is challenging for a gay or infertile couple who may not have the biological requirements to reproduce on their own and may not have financial or legal access to fertility treatment. Aside from what are often considered (by straight people) to be major life markers, queer time may resist violence done by other, more trivial social requirements; queer time, like other methods of time with identity-based qualifiers, is often jokingly meant to refer to being five to fifteen minutes late.

Interestingly, a significant amount of queer literature turns to Foucault or Nietzsche for queer conceptualizations of time; however, a considerable amount also turns to more modern embodied experiences like those described above: "The question of 'queer time' has become quite prominent in U.S. queer theoretical debates since the early 2000s. The most fundamental and consequential limitation of conceptions (and thus practices) of queer time to date is that they share with dominant heteronormative temporalities the assumption that time is ultimately linear—indeed, that it is 'straight.' Their intervention lies in slowing down, stopping, or reversing that linear trajectory, rather than calling it into question" (Boellstorff 2007, 229). While slowing down, stopping, and reversing are a good beginning to reconceptualizing time, they are all reactive to an initial linear conceptualization of time. Elizabeth Freeman (2007, 159) suggests that "we reimagine 'queer' as a set of possibilities produced out of temporal and historical difference, or see the manipulation of time as a way to produce both bodies and relationalities (or even nonrelationality)." In other words, queer temporality offers paths that are reactive to straight notions of time as well as paths that resist this reactivity.

Freeman (2007, 159) offers useful accountings of how this might work, arguing that "what has not entered the historical records, and what is not yet culturally legible, is often encountered in embodied, nonrational forms: as ghosts, scars, gods." She states that "temporality is a mode of implantation through which institutional forces come to seem like somatic facts. Schedules, calendars, time zones, and even wristwatches are ways to inculcate what the sociologist Eviatar Zerubavel calls 'hidden rhythms,' forms of temporal experience that seem natural to those whom they privilege" (160). To resist these structures is to potentially risk comfort and power, to refigure relations in ways that may be risky—ways that are queer.

In their roundtable discussion—a discussion articulated across time and whose time line was revised, according to Freeman (2007)— Carolyn Dinshaw, Lee Edelman, Roderick A. Ferguson, Carla Freccero, Elizabeth Freeman, J. Halberstam, Annamarie Jagose, Christopher S. Nealon, and Tan Hoang Nguyen articulate a variety of approaches to time, queer and otherwise. One common thread is a question that arises and re-arises about time's relationship to history. Freeman prompts, "Hearing you all talk, I'm wondering if it's possible to think relationality across time without some concept of history, and if history boils down to 'historicism'" (184). Dinshaw, referring to queer studies, says "some very searching theoretical work on history and historicism has appeared over the last fifteen years or so, but there's a tendency—at least among us literary scholars—to continue to critique 'history' (meaning old-style historicism) as if this work had never been done. Maybe this is an index of the difficulty of reworking linear temporality . . . figuring out the criteria by which different nonlinear temporalities might meaningfully be brought together—figuring out how to make heterogeneity analytically powerful—is exponentially harder" (186). In an echo of our earlier exploration of Baudrillard and Berlant, Nealon (in Dinshaw et al. 2007, 187) argues that "when we talk about time, we're talking about economies" (187). And Halberstam argues against reactivity: "I would argue for a kind of counterintuitive critique, one that works against the grain of the true, the good, and the right but one that nonetheless refuses to make a new orthodoxy out of negativity" (194).

Elizabeth Grosz (2004, 2) articulates "a reminder to social, political, and cultural theorists" that we too often forget the body—"the nature, the ontology, of the body, the conditions under which bodies are enculturated, psychologized, given identity, historical location, and agency." Grosz argues that our studies of subjectivity and embodiment lead us back to natural sciences and to questions we might have liked to avoid, but that we need to return to these conversations: "Time is neither fully 'present,' a thing in itself, nor is it a pure abstraction, a metaphysical assumption that can be ignored in everyday practice. It cannot be viewed directly, nor can it be eliminated from pragmatic consideration. It is a kind of evanescence that appears only at those moments when our expectations are (positively or negatively) surprised. We can think it only when we are jarred out of our immersion in its continuity, when something untimely disrupts our expectations" (5).

Grosz (2004, 5) further suggests that we think of time "in passing moments, through ruptures, nicks, cuts, in instances of dislocation," though time itself is continuous and oblivious to these human perceptions. Events that disrupt our immersion in the flow of time make up memorable emergences: "ruptures, nicks, which flow from causal connections in the past but . . . generate unpredictability and effect sometimes subtle but wide-ranging, unforeseeable transformations in the present and future" (8). Drawing heavily on Nietzsche and Darwin, Grosz shows that we tend to make assumptions about time and the way it works. For example, assumptions about the past might include that it is enough like the present and the future that we can learn lessons from the past to make better choices in the present or that the present must be understood as a contiguous effect of the past (114). Science as an enterprise has "been spectacularly successful in generating technologies that produce certain regularities, even if their explanatory power has left much of the universe—particularly its irregularities, the events that transform it, its qualitative metamorphoses,—unexplained or enigmatic" (245). That is, the scientific method—upon which are based technical processes such as clinical trials—ensures particular rigid conceptualizations of time and how it works toward replicability. However, the scientific method fails to account for complex processes that cannot be tested in a vacuum. For example, Colleen Derkatch (2016) demonstrates that clinical testing of acupuncture actually introduces flaws in the process, resulting in a test not of acupuncture as it is truly practiced—with emphasis on the client-patient relationship—but rather of a corporeal practice that is just a shadow of what an acupuncturist would ordinarily do. This is testing of the body as though it could be separated from the mind: "To understand material systems, including their ramifications in and as cultural systems, that is, in their interconnectedness, one cannot simply add together the various well-analyzed, independently conceived subsystems. Their integration is not additive but transformative; it is performed not only through spatial relations but above all in temporal relations" (Grosz 2004, 245). The *process* of acupuncture over time is impossible to map onto standard clinical testing procedures. Our conceptualizations of science, our reductionist approach to understanding, means that without expanding our sense of time, our sense of connectedness, our sense of what is real, some things become un-testable.

Perhaps, like science, feminisms have limited themselves with rigid conceptualizations of time; perhaps by reconsidering the temporal assumptions of our theoretical frameworks, we might open up new possibilities for understanding and action. José Esteban Muñoz (2009) connects this to the idea of resisting notions of objectivity, which I articulated as an important frame for the present book. "This book has been written in nothing like a vacuum" (15) he says about his book, arguing that "the refusal of empiricist historiography and its denouncement of utopian longing has been an important cue for this project" (17). For queer theory, notions of dystopia and utopia figure as structural parallels to a more individualistic sense of despair and hope, which arise from the dismal prospects for longevity and happiness for out queer people in recent history in the United States (Love 2007). Heather Love discusses a "turn to the negative in queer studies" associated with the popularity of Foucault (2). Her book focuses on figures that turn backward, looking back at history: "One may enter the mainstream on the condition that one breaks ties with all those who cannot make it—the nonwhite and the nonmonogamous, the poor and the genderdeviant, the fat, the disabled, the unemployed, the infected, and a host of unmentionable others" (10). Love uses affect as a way to bridge and characterize relations between past and present. She shows that queer scholars "have shifted the focus away from epistemological questions" and "made central" the ways affect sponsors relations across time (31): "Tarrying with this negativity is crucial; at the same time, the aim is to turn grief into grievance—to address the larger social structures, the regimes of domination, that are at the root of such pain" (151). In other words, Love focuses on the mobilizing potential of negative affect. This turn and the material reality it arises from insist that this book's borrowing of notions of queer temporality gives rise to something that can contribute back to queer theory; I address the need for reciprocity in more detail in chapter 4.

Sara Ahmed (2011, 160) expands on this idea, pointing out that "queer theorists have been the most vocal in refusing to affirm the future, refusing to embrace the future in a politics of affirmation." Drawing on queer work's embrace of pessimism/nihilism, she suggests that "happiness is interesting" but that it may not look like what we imagine. What effect, then, does the potential for queer optimism have on our understandings of time? According to Ahmed, "Perhaps the queer point would be to suggest that we don't have to choose between pessimism and optimism. We can explore the strange and perverse mixtures of hope and despair, of optimism and pessimism, with forms of politics that take as a starting point a critique of the world as it is and a belief that the world can be different" (161). Cheryl Glenn's (2018) politics of hope exists in conversational tension with Ahmed's claims; the two taken together can perhaps help us find a useful balancing point in thinking about conceptualizations of time and their impacts on female and queer bodies.

Turning to rhetoric, E. L. McCallum and Mikko Tuhkanen (2011, 8–9) argue that "temporality is necessarily already bound up in the queer. This temporality, we further suggest, is not that of *chronos*, or linear time whose very name mythically signals lineage (in the ancient Greek myth, Kronos is father to Zeus); rather, the contingencies of the queer might be closer to the time of *kairos*, the moment of opportunity." Meanwhile, Stockton (2011, 345) returns us to lived experience and reflects on rhythm, on the rhythm involved in the lifestyles of (it seems) queer academics: "Like me, you're likely to be uber-Protestant-work-ethic-hounds at your labors six days a week, with one day—one blessed day—for queer hedonism lived to the hilt." She says, "Queers . . . enjoy driving hard to the wall, in our work, to theorize the nonproductivity of luxury and the temporality of certain nonnecessities. We even fight politically on hunger, on health care, on the redistribution of wealth, so as to make our brand of destruction more available and temporally frequent" (348).

Elizabeth Freeman (2010, xi) asks how we might "think against the dominant arrangement of time and history." She theorizes chrononormativity as "the interlocking temporal schemes necessary for genealogies of descent and for the mundane workings of domestic life" (xxii). This approach has effects not just on how we understand events (like the Deepwater Horizon Disaster) but, perhaps more important, on how we understand the theories (apparent feminisms) and concepts (efficiency) that frame those events: "Looking backward, I can see how the crisscrossing energies of postcolonial studies, studies in medieval and other so-called premodern periods, and critical race theory made the questions of time's sexual politics (and the temporal politics of sex), if not inevitable, at least already asked in several different idioms" (xxiv). And importantly, "Time can produce new social relations and even new forms of justice that counter the chrononormative and chronobiopolitical" (10). What new forms of justice might different understandings of efficiency produce in relationship to feminist technical communication? What forms of justice do our current understandings of efficiency and time preclude? Such questions lead me to end this section with Freeman's conclusion: our obligation is to "jam *whatever* looks like the inevitable" (173, original emphasis). That is, anytime something appears to be a foregone conclusion, we should throw up red flags and re-examine the assumptions that make such a feeling possible.

Time is a useful lens through which to examine foregone conclusions because alterations in the scope of time might allow us to see how windows of normalcy shift over longer periods. One example of a temporally dependent ideologic reframing in the context of environmental concern and crisis might be recent work by Sylvia Jaworska (2018, 214), who offers evidence that rhetorical choices related to environment have significant impacts on public opinion and policy by showing that oil company documentation has participated in a shift from the more provocative and exigent "global warming" toward the more benign "climate change":

> Examining frequencies of "climate change" and synonymous terms reveal [*sic*] a strong preference for "climate change" and other terms such as "global warming" or "greenhouse effect" are rarely used in the studied sample. "Global warming" is a term associated with urgency and human agency, and seen as more threatening in contrast to "climate change," which is more general in meaning... The greater use of "climate change" and the near absence of "global warming" in the studied sample might indicate the preference of the oil industry to frame climate change as a more distant and less threating phenomenon.

In the same way Jaworska hypothesizes rhetorical distancing as a result of this shift in terms, I am suggesting that enacting a shift in terms here can result in the opposite sort of relocation—a relocation that brings into focus a crisis previously made distant, out of focus, circumspect. Nicole I. Caswell and Rebecca E. Johnson (2022), for example, frame systemic racism as a crisis in need of immediate redress (alongside Covid-19) in their university writing center study. Their work ultimately shows the necessity of community responses to public crises. Theorizing slow crisis helps us recognize a crisis that exists even if it does not exist on our usual time line for crises. The ramifications of our understandings of time on material orientations can be explicated, at least in part, by queer understandings of temporality. The varied approaches to temporality as a way of beginning to elucidate how rhetorical recognition affects cognition—and how the stakes of this recognition are rather high—that I have touched on here might be only the beginning. One direction they might send us off in, for now, is an exploration of the relationship between queer conceptions of temporality and complementary feminist scholarship.

4
DISASTER

The Deepwater Horizon event began on April 20, 2010, when an explosion onboard the Deepwater Horizon oil rig killed eleven men and opened the connected undersea well. The breached well then gushed oil into the Gulf of Mexico for several months until it was permanently plugged in mid-September of that year. British Petroleum PLC, an international corporation, was leasing the Deepwater Horizon rig at the time and was held largely responsible for the spill in the public eye, though many other actors were involved. The material effects of this disaster for those eleven men were immediate. The material effects for their families and colleagues were almost immediate. The material effects for the ecologies, economies, and people in Gulf Coast communities continue. Yet, when we think of oil spills, tankers, rigs, and corporations, we rarely think of feminism, with its focus on materiality. We need to do so.

The Deepwater Horizon Disaster (DHD) is, to my mind, a context that obviously stands to benefit from feminist analysis. However, I suspect that most people don't hear "Deepwater Horizon" and immediately think "feminisms." For me, one important connection is that the DHD could likely have been avoided altogether or significantly mitigated if feminist communication principles had been enacted at a number of points along the crisis time line. Further, connections between oil disasters and feminisms exist along the trajectories of two material truths: that water is a starting point for life much in the same ways women are starting points for life and that our understanding of the interaction between humans and ecosystems is fraught with false boundaries. (Humans are, after all, *part* of ecosystems.) My home institution, East Carolina University, recently opened a Water Resources Center that is dedicated to "finding ways to improve our understanding at the interface of human, natural and engineered systems to ensure a resilient and sustainable future for communities at home and around the world" (https://ecuwater.org/). This quote acknowledges the importance of communicative analyses; tellingly, it also points toward the blurring of what counts as human and what counts as natural.

For some people, the term *feminist technical communication* seems to be an oxymoron. Some perceptions of technical communication invoke a field so technical, so efficient, so objective, so prestigious, so *masculine*, that a feminist version of such a thing sounds impossible. The field of technical communication may seem to the public to be attached to "functionalist, skills-based approaches"; indeed, Godwin Y. Agboka and Natalia Matveeva (2018) document this sense (and resistance to it) from scholars in the field. It is my contention that this public sense of objective, functionalist technical communication is precisely why technical communication badly needs feminisms. My impetus for this project lies in my belief that technical communication—and world politics, for that matter—could benefit greatly from feminist perspectives. Feminists have long critiqued the tendency of hegemonic discourses to make unapparent the value of embodied epistemologies, and many feminist works have argued that experiential knowledge is both important and overlooked (Belenky, Clinchy, Goldberger, and Tarule 1986; Hesse-Biber and Leavy 2007; Ramazanoğlu and Holland 2009). As some short examples, even simple instruction manuals point to the importance of recognizing bodies when we are producing specialized documents, and the technomedical takeover of medicine was all about bodies and consolidating power with certain bodies. To explain—and to paraphrase a lot of historiographical work done by Barbara Ehrenrich and Deirdre English (2010)—women dominated the field of healing until enterprising men realized they could use technical procedures like licensure to criminalize experiential/embodied knowledges to elevate their own prestige. This mindset continues to make practices such as midwifery—or getting attention when one is among a limited number of corporate victims, such as people who developed skin rashes due to the Deepwater Horizon oil spill—difficult to sustain. In fact, prestige (or ethos) is often a motivating factor in ignoring bodies and embodied practices. If we can leave our bodies behind, we can stake a better claim to objectivity.

For this reason, many male-dominated professions are in desperate need of new epistemological approaches. Feminist approaches, in such contexts, can enliven fields where rhetorical patterns have become so fossilized that they are no longer questioned or innovated upon. One thing feminisms do very well, as noted above, is point to materiality. A recent resurgence in materialisms and feminist materialisms drives this point home (Booher and Jung 2018; Coole and Frost 2010), but for my purposes here we can begin with the idea that feminisms help us return to bodies. Many large systems work to separate bodies (particularly women's bodies) from technical information and statistics. For example, in the context of biomedicine:

> Only considering the "standard" data about the body eliminates the context necessary in understanding how diverse bodies respond to disease and illness and, as a result, how they might be diagnosed and treated. According to N. Katherine Hayles (1999), "Information, like humanity, cannot exist apart from the embodiment that brings it into being as a material entity in the world; and embodiment is always instantiated, local, and specific" (54). While data about bodies allow us to have a broader view and generalize among groups and categories of bodies, they can also be limiting and exclusionary if embodied experiences aren't also accounted for. (Frost and Eble 2020, 6)

Feminisms are concerned with the interconnectedness of embodied experience and systemic influence. They question, critique, and build on social constructions of gender and the material effects of those constructions. At the same time, they are attentive to the ways "the body itself seems an impossible object with which or through which to think historically. [Fredric] Jameson suggests that 'body theory' is actually the symptom of a certain loss of time itself" (Freeman 2010, 10). In other words, feminisms seek to attend to both big-picture system analysis and the truth and value of individual embodied experiences over time, as best they are able.

Along these lines, Jeffrey T. Grabill and W. Michele Simmons (1998) argued for a critical rhetoric of risk communication, suggesting that risk is socially constructed. They suggested that definitional disputes in risk communication situations "are a public contestation over the meaning of risk—the 'truth' about risk is actually a product of such disputes" (423). Grabill and Simmons advocated for explicit citizen participation in the construction of rhetorics of risk. They were, in part, taking up a call made by Alonzo Plough and Sheldon Krimsky (1990), who expressed the need for situating risk communication within a cultural framework. Plough and Krimsky asserted that "if the technosphere begins to appreciate and respect the logic of local culture toward risk events and if local culture has access to a demystified science, points of intersection will be possible" (229–230). In other words, this is an argument for moving technical communication rhetorics back to culture, back to embodiment. That move is one I seek to follow in this project.

In this chapter, I offer an apparent feminist telling of the Deepwater Horizon Disaster. I begin with a section outlining some of the reasons we may want to turn to materialism and be concerned with human health—with our bodies. That is, I get into detail related to the reasons for suspecting that the DHD would have impacts on human health. I then move into a mixed-methods and partly methodical, partly serendipitous history of gathering information. My approach to storying my information gathering springs from an apparent feminist impulse to showcase that the ways I, the researcher, encountered information about the DHD matter. The information-gathering section includes discussion of two research trips, insider knowledge of Gulf Coast economies, and textual research. This chapter is explicit about making choices to offer certain versions of these histories (table 4.1). It is also critical of some of the histories that have been constructed in the media.

Ultimately, this chapter takes up *efficiency* as a framing idea. Efficiency is the frame that requires that these histories be told in a particular way, and I am operating under a *new* efficiency frame—one that sees human health as an important contributing value to efficiency—to tell them differently. We understand efficiency in a certain way, and that has limited our possible responses to crises like the Deepwater Horizon Disaster. If we understood efficiency differently, we'd have different options. For example: efficiency overrides health concerns in a variety of contexts, particularly when slow crisis renders those health concerns less recognizable. Operating under a standard efficiency frame, we tend to respond only to human health concerns that can be framed as traditionally urgent. Under a revised efficiency frame, other/additional responses—like those in this book—become possible.

HUMAN HEALTH

In April 2014, I browsed an archive called *Engineering Case Studies Online*. This archive was organized by specific events, one of which was the Deepwater Horizon Disaster. Within the Deepwater case, I found a variety of documents. Some—like images archived in public sites including Wikipedia—have circulated widely. Others, most of which are accident reports with corporate or government authors, are public documents that have enjoyed a far more limited amount of what Jim Ridolfo and Dànielle Nicole DeVoss (2009) term *rhetorical velocity*. What I did not find in my time with this archive were documents directly concerned with the effect of the oil spill on human bodies. This is perhaps unsurprising, given that we're talking about an engineering archive; yet engineers, too, exist in bodies. In order to collect the data shown in the archive, someone had to physically collect it.

In fact, very little of my research on the Deepwater Horizon Disaster has unearthed documents that are directly concerned with the effects

Table 4.1. Selected, localized, and traditional time line of the Deepwater Horizon Disaster (all dates are 2010)

Date	Event
February 15	Transocean's Deepwater Horizon rig (which had previously drilled the deepest subsea oil well in history) begins drilling in the Macondo area.
April 17	Deepwater Horizon completes drilling weeks behind schedule.
April 20	An explosion onboard Deepwater Horizon kills eleven men, injures seventeen others, and precipitates concerns about leaking oil.
April 21	Oil is observed on the surface of the water up to two miles from the rig, though much of the leaking oil has burned.
April 22	Deepwater Horizon sinks. Leaks are confirmed. The Environmental Protection Agency (EPA) authorizes use of surface dispersants.
April 29	State of emergency declared in Louisiana as oil approaches coast.
April 30	Oil washes ashore in Louisiana.
May 5	One of the smaller leaks is capped.
May 9	Tar balls wash ashore in Dauphin Island, Alabama.
May 11	BP, Halliburton, and Transocean officials testify before the US Congress. They disagree on both blame and the amount of oil being leaked.
May 15	The US Coast Guard and the EPA authorize the use of dispersants underwater.
May 23	BP and the EPA tangle over use of dispersants (Corexit).
May 29	Top kill procedure (which aimed to seal the well by pumping drilling mud into it) fails.
June 4	Oil washes ashore in Florida.
July 22–24	Containment efforts are abandoned in fear of Tropical Storm Bonnie.
September 19	Macondo well is permanently sealed four months after the leak began.

of the oil spill on bodies. I'm not alleging evidence of malice; I *am* interested in why the two divergent narratives of oil spill recovery focus on economics or ecology but don't often recognize human bodies as significant actors in either of those systems. The lack of attention is unsettling, since effects on human health from oil spills have been well documented. As reported in the *Guardian*, "History has shown responders to oil spills often suffer headaches and other symptoms, and in the long term are at higher risk of central nervous system damage, kidney and liver damage, and cancer . . . In addition, US worker safety regulations do not apply more than three miles offshore, leaving workers based near the ruptured well exposed" (Goldenberg 2010).

Bradley S. King and John D. Gibbins (2011), in a health hazard evaluation of Deepwater Horizon response workers, identify several types of direct offshore response work—chemical dispersant application (aerial and surface), in-situ burning, booming, skimming, and vacuuming—as well as several types of onshore response work: wildlife cleanup, beach cleanup, and decontamination and waste management. Chemical dispersant application involved Corexit® EC9500A and 9527A dispersant (notably, Corexit is banned in the United Kingdom, where BP is based) released into the water from maritime vessels and aircraft, although this report evaluated only workers on maritime vessels. The Center for Biological Diversity (CBD) (2020) notes that Corexit 9527A contains 2-Butoxyethanol, which is known to harm red blood cells (hemolysis), the kidneys, and the liver. (The CBD reports that cleanup workers on the Exxon Valdez spill "suffered health problems afterward, including blood in their urine as well as kidney and liver disorders, attributed to 2-Butoxyethanol.")

Dispersants do not reduce the amount of oil but rather break down oil into smaller units. In-situ burning is a technique where workers intentionally light an oil slick on fire to burn it away. Booming, skimming, and vacuuming are all mechanical means of controlling oil slicks (figure 4.1). Wildlife cleanup most commonly involved cleaning oil off of birds. Beach cleanup involved mitigating oil-soaked sand and cleaning up tar balls; it included work on twenty-four sites with "light residue," six with "moderate residue," and three with "heavy residue" (King and Gibbons 2011, 7). Decontamination and waste management involved sanitizing boats and other equipment used in the cleanup.

Each of these types of response work requires some level of human exposure to oil, chemical dispersants, or both. King and Gibbins (2011) begin their report with the May 26, 2010, hospitalizations of seven fishermen who were involved in oil spill cleanup. These hospitalizations were followed by ten more. Details on the patients' exposure to oil, dispersants, or both are extremely limited. King and Gibbins ultimately conclude that none of these hospitalizations can be definitely linked to exposure to oil or dispersants—citing instead work-related concerns like fatigue and heat—although two of the workers did receive medical advice to avoid continued exposure. In the section on beach cleanup, King and Gibbons write that they "saw no evidence of exposure to dispersant at the shore cleaning sites"—a curious claim, as it is unaccompanied by any explanation of the means by which onshore dispersant exposure would be observable (7).

Later, in a section on in-situ burning, King and Gibbons (2011, 4) write:

> During the evaluation, we conducted PBZ [personal breathing zone] and area air sampling on shrimping trawlers towing booms during in-situ burns and on boats from which the burns were ignited. Sampling was conducted for VOCs [volatile organic compounds], aldehydes, CO, H_2S,

Figure 4.1. Response boats distribute boom to clean up oil where the Deepwater Horizon rig sank in 2010 in this image from the Defense Visual Information Distribution Service.

benzene soluble fraction of total particulate matter, diesel exhaust, and mercury. Exposures for all compounds sampled were either below detectable concentrations or well below applicable OELs [occupational exposure limits], with one exception being a peak exposure of 220 parts per million (ppm) of CO recorded on the double-engine ignition boat. This peak was likely due to the build-up of exhaust from the gasoline powered engines when idling with no movement of the boat and little wind.

While their hypothesis seems valid, its inclusion is a bit of an oddity since the source of a chemical that is exceeding an occupational exposure limit is irrelevant to its effects on human bodies. That is, the carbon monoxide will still affect the fishermen who've been pressed into service regardless of where it came from. For reference, 2012 Occupational Safety and Health Administration standards state that the permissible exposure limit over an eight-hour period is 50 ppm—less than a quarter of the exposure recorded by King and Gibbins—and also that the eight-hour rule has exceptions: "Maritime workers, however, must be removed from exposure if the CO concentration in the atmosphere exceeds 100 ppm." In other words, the peak exposure that King and Gibbins quickly explain away was a significant exposure, even if short in duration (as it might be if due to exhaust buildup).

Response workers were likely the people most at risk of exposure to oil and cleanup-related chemicals, but they were not the only ones with potential exposure. Two other significant categories of people exposed were those who live and work in proximity to the Gulf and beachgoers. While tourism took a significant hit in the aftermath of the explosion,

Figure 4.2. This hastily constructed berm was a new feature on Dauphin Island just in advance of oil coming ashore in June 2010.

beach traffic was not eliminated and locals were not evacuated or otherwise barred from potential exposure. The only significant regulation, in fact, was a temporary fishing ban (meanwhile, fishers were hired as cleanup workers). At the time of the oil spill, it was possible to find blog posts, particularly on the now defunct GulfofMexicoOilSpillBlog.com, that included complaints of side effects like headaches and images of people with skin rashes ostensibly due to oil (or dispersant) exposure. These posts and images are now mostly gone—although some are still viewable on public web archives and in an archived HuffPost Contributor page (Ott 2010).

INFORMATION GATHERING

In what follows, I first take the reader sequentially through two research trips to the Gulf Coast conducted in 2010 and 2013, respectively. In 2010, I visited the small Alabama community of Dauphin Island as the effects of the Deepwater Horizon Disaster were still very much apparent in national media (figure 4.2). In 2013, I visited Mobile, Alabama, to investigate what continued effects the disaster had had on the local economy, ecology, and human health. Then, I offer an analysis of texts related to the DHD to further contextualize my findings. The textual analysis focuses on the prevalence of DHD information on government health sites as well as on two websites (one related to economic harms and the other to health risks) that explain class-action lawsuits related to the Deepwater Horizon Disaster.

Dauphin Island Research Trip, 2010
My findings during this initial research trip rely on my experiential knowledge of the small community of Dauphin Island, Alabama. My data collection method on this trip was one of both necessity and fortune; as a regular and longtime but simultaneously temporary member of the Dauphin Island community, I collected data in whatever ways were accessible to me at the time of my visit. I did not go to Dauphin Island in the summer of 2010 intending to study transcultural risk communication; I went to spend time with my family as we had done for many years and, if possible, to aid in the cleanup effort.[1] I experienced the artifacts I describe here first as a citizen rather than as a researcher. The same status that obligates me to implicate myself also allows me a perspective on these artifacts that would not otherwise have been possible. When I discuss this time period as a research trip, it is important to understand that it began as something else and became a research trip somewhere in the middle of my time there.

The Deepwater Horizon oil spill had—and continues to have—far-reaching effects, but Gulf Coast community members were the people most immediately and deeply impacted. Dauphin Island, Alabama, is a Gulf Coast community of a little more than 1,000 residents, though the actual population fluctuates greatly with seasonal tourism. The name "Dauphin Island" references both an incorporated town and the small island—less than 2 miles wide (3.2 km) and 14 miles (22.5 km) long—on which the town exists. The island borders the southwest side of Mobile Bay and is connected to the mainland by a 3-mile-long (4.8-km) bridge; this location means Dauphin Islanders are no strangers to disaster or to disaster rhetorics, having weathered numerous hurricanes. My family vacationed on Dauphin Island for years, and—although I still think of myself as a visitor and tourist—I developed a sense of belonging to and responsibility toward the community. In 2010, I stayed on Dauphin Island from Saturday, June 19, through Friday, June 25. During the time leading up to my trip and during the duration of my stay, the Deepwater Horizon well had not yet been sealed; indeed, several efforts to seal it had failed and the general feeling in many Gulf communities was one of extreme frustration.

My focus during this research trip and in the work I did after it (Frost 2013b) was on the ironic locations of different communicative acts related to the DHD. To explain, between April and September, as the disaster continued to unfold, some international and national communicators—including BP and the US federal government—created documentation that suggested primarily economic understandings of the risks and dangers associated with incoming oil for Gulf Coast communities. Meanwhile, local Gulf Coast sources engaged in disaster response rhetorics that focused on ecologic as well as economic risk. This importance of ecological understandings of risk in local rhetorics—and the lack thereof in international and national discourses—highlights the differing goals of specific communicators in shaping understandings of risk associated with the Deepwater Horizon Disaster. It also highlights what is absent—any focus whatsoever on human health.

Specifically, my interest in 2010 focused on the following artifacts:

- A BP-authored public notice posted at the Dauphin Island Marina designed to recruit local boat captains to work for BP as part of the cleanup
- Two informational bulletins posted in community centers and authored by federal agencies (These artifacts explained to citizens how to apply for compensation for lost wages and other damages due to the oil spill. The bulletins were sized to fill the space of the previously existing community bulletin boards, meaning they were approximately 3 feet [0.9 m] tall by 5 feet [1.5 m] wide.)
- The website of the Dauphin Island Sea Lab (DISL), a regional organization with ties to higher education (Dauphin Island Sea Lab 2010)
- A blog published by the Dauphin Island Real Estate (DIRE) company (2010)
- The website of the Town of Dauphin Island (2010).

Prior to my trip, I used online media to track the effect of the oil spill on the Dauphin Island community. Specifically, I found information in a blog published by Dauphin Island Real Estate, Inc. (2010), on the Dauphin Island Sea Lab's (2010) website, and on a website hosted by the Town of Dauphin Island (2010). These sites were produced by people whose livelihoods depended on the health of the Dauphin Island ecosystem and economy. These digital spaces were largely concerned with defining and mitigating the potential risks the oil spill had created for the specific area of coastal Alabama, and they utilized a number of rhetorical techniques to construct that risk in more positive ways for local interests.

I found that sources written by people who lived on Dauphin Island—local communicators—were overtly persuasive and much more helpful in determining the likelihood of oil coming ashore at Dauphin Island than were works by professional communicators at major media outlets. For example, the Dauphin Island Real Estate (2010) blog posted frequent local beach updates, whereas global communicators focused only on areas where oil was already coming ashore. Nevertheless, when I arrived on the island, I found few non-oral local rhetorics evidenced in public. Instead, I found several written, public instances of risk communication by federal and international writers. These artifacts on and near the island were highly technical in nature in that they used the genre conventions—such as small type and official seals—of documents that are widely considered to be objective, although technical communicators may often consider them culturally saturated (Haas 2011). For example, BP placed literature at the Dauphin Island Marina; the literature was designed to recruit local boat captains to work for BP as part of the cleanup. At two other locations, informational bulletins posted on preexisting bulletin boards explained to citizens how to apply for compensation for lost wages. One of these bulletins was posted at Dauphin Island's only gas station, the Chevron station on Bienville Boulevard. Its central location and its monopoly on gas make the Chevron station a community center; the space this bulletin came to occupy was so culturally saturated—it had previously been a space to post local advertisements to sell boats, publicize repair services, and so on—that anything posted there would command only one small sliver of the many layers of cultural meaning already attributed to that space. The other bulletin I found was posted just outside the front door of the Pelican Reef restaurant in Theodore, Alabama, about 25 miles (40.2 km) north of Dauphin Island. This bulletin, identical in nature to the one at the Chevron station, evidenced a more regional application of this type of rhetoric. I mention this bulletin to show that federal response, at least within the region, was standardized.

I have argued (Frost 2013b) that the delivery methods of these artifacts are ironic in that international and national entities utilized local spaces whereas regional and local communicators turned to globalized digital sites to narrate the disaster. However, it is worth noting that they narrated different versions of it. Specifically, government- and BP-authored documents that were inserted into physical spaces on Dauphin Island sponsored largely economic understandings of the effects of the spreading oil; regional and local sources engaged in more nuanced constructions of combined and interdependent economic and ecologic risk, if not of risk to human health. This scenario is evidence of an obligation by professional and technical communicators to pay attention to complex transcultural flows of communication that move between local and global cultural spaces as they participate in constructing both the risks and the histories of particular events.

Mobile Research Trip, 2013

In late November 2013—three-and-a-half years after the explosion onboard Deepwater Horizon—I traveled to the Gulf Coast for on-site research on health services for those affected by the Deepwater Horizon Disaster. I wanted to know how the public health department in Mobile County was talking to citizens about the effects of the spill. More specifically, I wanted to answer the research question: what scientific data sets inform(ed) the action plans of healthcare facilities with responsibilities to citizens of the Gulf Coast in the aftermath of the Deepwater Horizon oil spill? My training as an investigative reporter suggested to me that it would be best if the health department did not have notice of my arrival; I wanted to know what conversations were happening about health, bodies, and oil when no one was watching. I planned carefully for the trip and hoped to distribute surveys to adult users of the local public healthcare system and to conduct interviews with administrators of that system. Most important, I wanted to spend my time in Mobile gaining an understanding of the interlocking systems that helped the city run throughout the DHD. While I have experience in more rural economies in general as well as in this area,[2] I had previously spent very little time in Gulf Coast cities.

My relative inexperience in urban settings became apparent as soon as I entered the main building of the Mobile County Health Department (MCHD). I had imagined a lobby or waiting area where I could leave and distribute surveys; instead, the building's architecture forced visitors to choose a particular specialized area before providing anything resembling a lobby or gathering area. Since such specialized areas would draw particular populations and I had hoped to distribute my surveys to a cross-section of public health users, this immediately threatened to skew my data. I determined to leave surveys at every lobby in the building and was directed to the Human Resources Department to obtain permission to do so. (The surveys, which I never distributed, asked if each participant had any health concerns related to the oil spill, if they had received any information on oil spills and human health, and how helpful or unhelpful that information had been.)

A conversation with the director of human resources was disheartening. The director promised to check with her superiors about whether they might allow me to distribute surveys. Intuition developed from my time as a reporter let me know that this was an empty promise—which did, indeed, turn out to be the case, as the callback never happened despite repeated follow-ups on my part. I set about asking my interview questions, beginning with whether the MCHD had produced any communications related to the Deepwater Horizon Disaster, but the director told me that I needed to speak with their environmental health specialist. When I asked how to contact that person, I was informed that she was on leave through mid-January. I was further informed that I could not interview administrators without permission from "higher up," which I was also promised a callback about. Unsurprisingly, that call never came through, either.

I have never constructed moments like this as research failures. Rather, I found exactly what I was looking for: I learned that public health services for those who were concerned that they may have been affected by the Deepwater Horizon Disaster were largely inaccessible, and public healthcare workers in Mobile County have little to offer citizens concerned about the effects of the spill on their bodies. It is unlikely that a person who came to the MCHD with concerns about their health related to the oil spill and without university credentials would have gotten any farther than I did, especially given my relative privilege in this situation as an educated white woman. In particular, the circumstance of the environmental health specialist being on leave for the rest of the year with no replacement would hardly be helpful to someone with immediate concerns about the environment's impact on their health, and this certainly represents a failure by the MCHD. At many points during this project, I thought back to my career as a journalist and considered how I might have presented the information I'd found—and not found. The lede I most often thought of during this research trip sounded something like this: "People searching for answers about healthcare related to the oil spill are on their own."

After encountering these roadblocks on my initial foray into seeking information through the MCHD, I spent most of my time in Mobile County observing public healthcare locations, collecting literature, and paying attention to what local people had to say about the Deepwater Horizon Disaster. In the interest of due diligence and learning about access across Mobile County, I visited all seven non-specialized MCHD locations in Mobile County:[3]

- 251 North Bayou Street, Mobile, Alabama 36603
  This facility includes the MCHD main office and a WIC [Special Supplemental Nutrition Program for Women, Infants, and Children] clinic. It's in a large building with no main waiting area. I was not able to locate any literature available to the public.
- 248 Cox Street, Mobile, Alabama 36604
  This facility includes the Newburn Clinic, TEEN Center, and Women's Center. A small brochure center was available with several

pamphlets, most on HIV/AIDS and all available in Spanish and English. There was no main waiting area.

- 5580 Inn Road, Mobile, Alabama 36619

    This facility appeared to be brand new and had signage indicating as much. Most clients seemed to be traditional nuclear families (or at least appeared in groups of one man, one woman, and one or more small children). There was a main waiting area, but I was not able to locate any literature.

- 3810 Wulff Road, East Semmes, Alabama 36575

    This clinic appeared to serve mostly traditional nuclear families, as described above. The receptionist said they did not have any information on environmental health issues.

- 4547 St. Stephens Road, Eight Mile, Alabama 36613

    This clinic focused on pediatrics. I picked up a brochure on pesticides and one on immunizations. A receptionist brought me several more brochures from the back, including literature on safe sex from StayTeen.org, lead poisoning, abstinence, HIV, allergies, and a survey on teen pregnancy rates.

- 19250 N. Mobile Street, Citronelle, Alabama 36522

    This clinic appeared to be very new. The receptionist was busy with a patient when I came in and did not acknowledge my presence. I waited for half an hour and then left. There was no health literature available in the waiting room.

- 950 Coy Smith Highway, Mt. Vernon, Alabama 36560

    At this clinic, the receptionist said someone else had recently come through asking for literature on environmental health issues. She gave me a brochure on area clinics.

At every location, I explained that I was looking for information on environmental health, especially anything related to the oil spill. Only one location was able to offer any information on environmental health. At the Eight Mile clinic, I found brochures on pesticides; in addition, a receptionist brought me several more brochures from the back, including literature on lead poisoning and allergies. These brochures were, thus, not readily available to the public but presumably would have been made available to anyone with queries similar to mine. The pesticide brochure came the closest to addressing the sorts of questions a person might have about the human health effects of an oil spill, with language such as: "Is your neighbor using pesticides, and you're worried about possible risks" and "Our 'user-friendly' scientists will help you . . . assess toxicity and risks." These are the sorts of questions a person might ask about exposure to oil or oil dispersants, but nowhere in the literature was mention made of contaminants other than pesticides. Much of this literature was addressed to parents about risks to their children; the receptionist told me that this particular location had a special focus on pediatrics, although that information was not evident outside the clinic or on any public communication about the clinic that I could see.

At the last clinic I visited, in Mount Vernon, Alabama, the receptionist told me she'd been "alerted" that someone who was working on a project would come through and that she'd been told to give anyone doing such work a brochure and not to agree to an interview. The brochure she handed me listed the area clinics I'd already visited. This sort of circle-the-wagon reaction seemed odd given the relatively benign nature of my questions. I think it speaks to how few people ever ask questions about the environmental health of the MCPD, and it may also evidence some fear of legal repercussions. At this point in the trip, I was kicking myself for not having called ahead so the MCPD folks felt less defensive; however, this was a purposeful choice I'd made based on my work as a reporter, when it was almost always better to catch someone before they'd had a chance to practice their pitch, so you ended up with more authentic conversations and truthful information.

On this research trip, I also revisited three locations I'd focused on during my previous research trip: the Dauphin Island Marina, the Dauphin Island Chevron station, and the Pelican Reef restaurant in Theodore, Alabama. I found no evidence of literature related to Deepwater at the marina, which was nearly deserted. (Even the bait shop was closed, despite the fact that the hours indicated it would be open. This was around 3:00 p.m. on November 21, 2013.) The Chevron station had been remodeled, and the community bulletin space was gone. The community bulletin board at the Pelican Reef remained, but it was mostly empty. I drove the length of the island and observed that the outer berm had disappeared, although the inner berm (the one that's always been there) still existed. (A berm is a hill of sand intended to protect the island from incoming waves, water, and so on.) I saw one bulldozer on the beach, although I could not be sure if it was for moving sand or if it was part of home construction. I drove to the west-end beach and saw little evidence of anything pertaining to the oil spill.

I also visited the Dauphin Island Sea Lab (DISL). Outside, I found information about an oil rig just off the west end of the island. Inside the DISL (which hosts the George F. Crozier Estuarium), I eventually spoke with a senior aquarist[4] who showed me a display on Deepwater Horizon, and we talked about the efforts taken in the display to be "objective," in their words. We looked especially at a panel that talked about the importance of oil to the local economy. The aquarist was not part of designing the display, although they have a master's degree in ocean biochemistry. The aquarist refers to themself now as a "fishkeeper." The display read right to left, which seemed unusual and prompted a discussion of visual space. At this point, the aquarist returned to our conversation about objectivity in language and led me around the wall the display was attached to. On the other side of the wall, near the front of the estuarium, was a wall featuring its sponsors—where BP and Chevron logos were prominently displayed.

Finally, drawing on my reporter toolkit, I decided to go right to the people. As I posed my research question to locals at restaurants, museums, bars, and other downtown establishments, the consistent answer I received was that there is lots of information available about the economic effects of the spill and some information on ecological effects but not so much about health effects and very little about how any of these things might be connected.

Toward the end of this research trip, I began thinking about how my research question had to be shifted because it was the wrong question altogether. The data sets that were relevant to most public agencies in Mobile County had to do with money; they were the data sets that showed how the oil affected the fishing and shipbuilding enterprises that are the backbone of the Mobile-area economy. The data I had come to collect about human health did not exist. I then began asking:

- Why is there so little focus on embodiment in relation to this environmental disaster?
- In what relatively few places have concerns about health been aired, and what have been the reactions to these concerns?
- Why do people seem to be so much more concerned about the economy and ecology than about health? How can technical communicators help make the connections among these three things more apparent?
- How do these cultural impulses (to privilege economy/ecology) literally shape our bodies?

To answer questions like this, we need to pay attention to embodiment—even when such attention seems to run counter to existing cultural practices and conversations. This conflict—this tension between bodies and culture—was not something I expected to be working with in this case, and I think it leads to an important recommendation for technical communicators and apparent feminists: we must emphasize the difficulty of navigating moral conduct and cultural relativity and recognize the interconnectedness of economic and embodied concerns. In other words, despite the fact that the data I sought did not exist, I did not see this as a reason to abandon all attention to human health. Rather, I opened myself to related questions, particularly those also having to do with material consequences of the DHD.

DIGITAL SOURCES

A variety of digital spaces exist to narrate and construct the Deepwater Horizon Disaster. I discussed above the ways some sources local to Dauphin Island used websites to communicate about the DHD in the months after the initial explosion. In this section, I move past what we would recognize as the traditional point of crisis to focus on websites ostensibly aimed at helping people recover from the effects of the oil spill. I focus on people here not because ecologies are unimportant but rather because (as shown above) the human health facet of the disaster has been largely overlooked. Object-oriented ontologies might offer unique approaches for a project like this; however, I see a human-centered approach as the most vital and the most efficient (in the revised sense of the term) given my own methodological leanings. This does not mean I avoid mention of ecologies but rather that I always see them in connection with humans, whether that means thinking about the reciprocal effects of ecology and human activity or about the ways humans perceive ecologies.

I begin with the very first mention of the human health effects of the Deepwater Horizon Disaster that I was able to find online. That page was hosted by the Centers for Disease Control and Prevention, and it appeared in mid-June 2010. (Oil had been confirmed as washing ashore by April 30, leaving an approximately forty-five-day gap before any information was available.) The page in question focused on health effects for embryos, fetuses, and pregnant women and operated as a sort of quick fact sheet. The page read:

> Although the oil may contain some chemicals that could cause harm to an unborn baby under some conditions, the CDC has reviewed sampling data from the EPA and feels that the levels of these chemicals are well below the level that could generally cause harm to pregnant women or their unborn babies. The effects that chemicals might have on a pregnant woman and her unborn baby would depend on many things: how the mother came into contact with the oil, how long she was in contact with it, how often she came into contact with it, and the overall health of the mother and her baby.
>
> People, including pregnant women, can be exposed to these chemicals by breathing them (air), by swallowing them (water, food), or by touching them (skin). If possible, everyone, including pregnant women, should

avoid the oil and spill-affected areas. Generally, a pregnant woman will see or smell the chemicals in oil before those chemicals can hurt her or the baby. The EPA and CDC are working together to continue monitoring the levels of oil in the environment. If we begin to find levels that are more likely to be harmful, we will tell the public. For up-to-date information on monitoring data along the Gulf Coast, please visit EPA's website. (Centers for Disease Control and Prevention 2010)

This page is no longer available on the CDC site, but I have quoted the full version as it appeared on June 24, 2010. This page remains the only real acknowledgment of embodied effects on humans I've found, and aside from the fact that a web page was only accessible to the 70-something percent of the US population with the discretionary income to have internet access, the timing in which it appeared—about six weeks (an entire half trimester) after the disaster began—was less than ideal.

The next two sites of import here, both originally published in May 2012, operate as a set. The "Deepwater Horizon Medical Benefits" site (Deepwater Horizon Medical Benefits Claims Administrator 2012) and the (now defunct) "Deepwater Horizon Claims Center: Economic and Property Damage Claims" (Deepwater Horizon Economic Claims Center 2012) were explicitly linked in their respective main navigation. The Medical Benefits site is one of very few mainstream artifacts that acknowledges a connection between the oil spill and human health, although it does tend to focus on those who worked directly with dispersants; filing a suit does not ensure that effects on health will be officially recognized or that the suit will be successful.[5] The Economic and Property Damage site lists the following as acceptable losses: seafood, individual economic, individual periodic vendor or festival vendor, business economic, startup business economic, failed business economic, coastal real property, wetlands real property, real property sales, subsistence, Vessels of Opportunity charter payment, and vessel physical damage. This site was the first return when "Deepwater Horizon" was entered into Google in late 2014, suggesting it was getting a lot of traffic at that time.

Another early response—this one with a much more limited audience—to concerns about the oil spill and human health came from Women's E-News, a nonprofit news service focused on women's issues and women's perspectives on public policy that was originally conceived in 1996 and founded in its current form in 2002. In a July 22 article (which linked to the aforementioned CDC site), author Diane Loupe (2010) and her interviewees directed attention to something the CDC totally ignored: the effects not just of oil but of the chemical dispersants used to respond to the oil spill. Loupe's important story reported a number of other relevant facts that were not disseminated through mainstream media, including:

- The CDC said it was "unlikely" that people in coastal areas would come in contact with dispersants.
- The CDC also noted that dispersants can cause dry skin, respiratory irritation, eye irritation, pneumonitis, harm to fetuses, and worse side effects with prolonged exposure. Meanwhile, a senior scientist at the Natural Resources Defense Council reported that components in oil can cause acute health effects and have been linked to cancer. The EPA reported that humans might experience nausea as a result of breathing pollutants associated with the disaster.
- Many healthcare providers were unaware of the CDC recommendations regarding the oil spill.
- More than 800 people contacted US Poison Control Centers about exposure to potentially toxic substances related to the DHD.

As is perhaps obvious from the second bullet above, those with worries about exposure received different information from different sources. Further, most of this information was not widely distributed through mainstream media outlets or healthcare centers. Finally, importantly, the last bullet above demonstrates that people were concerned about exposure—which makes my inability to locate information related to human health and the DHD on my 2013 research trip troubling.

Other pertinent digital textual sources include regulations put in place since the Deepwater Horizon Disaster. For example, the Bureau of Safety and Environmental Enforcement (BSEE) Well Control Rule (2019), announced in May 2019, removes offshore drilling safety regulations put into place after a six-year study precipitated by the DHD. The new rule rolls back the requirement for government officials to complete inspection, allowing third-party companies to do this work. It also extends the time between inspections and removes a requirement for companies to alert the BSEE of false alarms. As expected, environmental groups and safety advocates oppose the Well Control Rule. Regardless, the connection between these sources and public understandings of human health related to oil spills is, frankly, minimal.

SHIFTING EFFICIENCIES

Altogether, the information I've gathered about the DHD and responses to it paints a picture of corporate efficiency. Parties that bear responsibility for the spill looked to a standard definition of efficiency—(the appearance of) the most result for the most people—as their value system for response. This has meant a focus primarily on economy and secondarily on ecology. Human health is an easier context from which to shift blame, in that it is possible to argue that individuals who encounter hazards (e.g., oil, dispersant) bodily in some way made a choice to do so, whether it was a choice to live or travel in a certain area or a choice to become a cleanup worker. As Rickie Solinger (2001) has adroitly pointed out, though, all choices are constrained.[6] Further, health and ecology and economy are intertwined. A fisher who chooses to become a cleanup worker and then faces health effects due to dispersant exposure had the "choice" not to participate in the cleanup—even though we know the temporary inability to make a living at fishing due to the spill may have made this decision not much of a choice. The appearance of choice and the appearance of offering a "result" for everyone involved play to an efficiency model that is familiar to us all. In this way, a rhetoric of efficiency came to obscure human health in the aftermath of the DHD.

The account I've given of the DHD in this chapter is a selected, localized one, as noted in the title of the time line I include. Imagine, for example, what a time line might look like if written by an individual living on the Gulf Coast. It would be more localized; they might note many things I did not, such as the date the federal government banned fishing, the date the beaches started to empty of tourists, the dates local states of emergency were declared, the dates when local fishers began to elect to work cleanup, the date when a rash first appeared on their skin. Such an account would be very different from the one I include but also still very traditional chronologically. We might also imagine a disaster time line with slow crisis as its underlying value, a time line that sees efficiency as a longer-term project. That time line might include a history of the Macondo area, the dates on which BP made strategic decisions leading to the deepwater rig being in the Gulf, the dates of legislation that enabled offshore drilling near Louisiana. In chapter 6, I offer samples of what such histories—those that are attentive to slow crisis, human health and rights, and longer-term efficiencies—might look like.

First, however, I offer a more detailed accounting of the history I've given above. Chapter 5 creates an additional layer of analysis that helps to demonstrate the development of apparent feminisms as a methodology for technical communication, connect transcultural and intersectional rhetoric, and make underlying efficiency models more apparent. To be clear, my own understanding of efficiency shifted over the time I gathered information about the Deepwater Horizon Disaster.

Both the information gathering and the analysis I offer in this book are partial and imperfect; knowing this, I make efforts to make my biases apparent and to look for places to wed complementary approaches. In particular, in this context, I see transcultural analysis as a useful tool for thinking about the confluence of economic impact, ecological effects, and human health. In other words, this three-part approach viewed through a transcultural lens helps us think about (one iteration of) feminist intersectionality.

5
AN APPARENT FEMINIST ANALYSIS OF THE DEEPWATER HORIZON DISASTER

A MAPPING OF RESPONSIBILITIES

While the story I tell above is always already from an apparent feminist perspective—being that it is made up of the particular histories I have chosen, as seen from my point of view and with an eye to material effects for marginalized subjects—more explicit apparent feminist analysis can help us see some of the ways the DHD unfolded given the efficiency rhetorics that have been socially privileged. Thus, this more explicit analysis can also help us see how things could have unfolded differently with different efficiency models in place. In this chapter, I engage in a detailed analysis of the information given in earlier chapters; readers might think of this as an additional layer of meaning for the artifacts and experiences I have described. In so doing—in following a structured apparent feminist approach—I more clearly make explicit my own feminist perspective; I look for complementary viewpoints from across cultural contexts (especially noteworthy here is Huiling Ding's [2009] notion of guerrilla media, which calls forth notions of apparency in relationship with perspective); and I question what efficiencies have allowed the prevalence of various perspectives that make their way into this narrative.

To organize this analysis, I first examine non-local/professional communication, then local and regional citizen communication, then guerrilla media (to complicate the aforementioned categories), and finally transcultural communication. Transcultural communication is a way of getting at intersectionality, as it allows us to see across (some, limited) cultural divides and to interrogate the efficiencies that undergird differing perspectives on the DHD. In other words, transcultural work helps us uncover what has been rendered unapparent: the vital connections among human health, economy, and ecology. As I move through these (sometimes overlapping) categories, readers will see that each framing responds to a different sense of efficiency, and each thus offers us different output—different takeaways—as a result. In sum, this chapter aims to show how the efficiencies we choose reverberate through all of our thinking on a subject, and it juxtaposes the various audiences and communicative structures feminisms might imagine themselves responsible for.

Non-local/Professional Communication

The three artifacts I focused on in my first visit to Dauphin Island—the marina advertisement and the two bulletins—can be seen as pieces of non-local communication and certainly can be categorized as professional communication. I label them in this way because although authors' names are not attached to such genres, they were almost certainly produced by people who are professional writers of technical documentation. Although the local artifacts I discuss in the next section are also examples of technical communication—and some of them are fairly sophisticated—they were likely written by people whose jobs are not defined by the task of writing. We must acknowledge this difference because of (1) the experience (and thus level of responsibility) implied by the term *professional*, (2) the differences in the writers' cultural backgrounds, and (3) the power associated with writing that is widely acknowledged as professional in nature. Also of import is the fact that the disaster unfolded very differently for local and non-local actors. Harking back to the notion of slow crisis, non-local actors did/do not have to live with the everyday effects of a disaster that has been unfolding and continues to unfold over generations; local actors, in contrast, are steeped in the knowledges and effects of slow crisis. More traditional takes on technical communication would likely consider these lived knowledges as *bias*; apparent feminisms frame them as experiential, and often intersectional, knowledges. In other words, despite traditional notions of ethos that might suggest it is associated with professional status and corporate culture, local actors have a more complex and sophisticated understanding of the (slow) crisis of the oil spill. Further, local communicators were able to use layers of meaning to establish their ethos in these digital spaces. The non-local or professional artifacts discussed in this section, however, carry with them a particular kind of status because of the historical, cultural associations of traditional professionalism.[1]

The Dauphin Island Marina advertisement was the only example of transcultural communication from an international source that I found on the island. This artifact directly represents the interests of BP in Dauphin Island. The marina is a local gathering point as well as

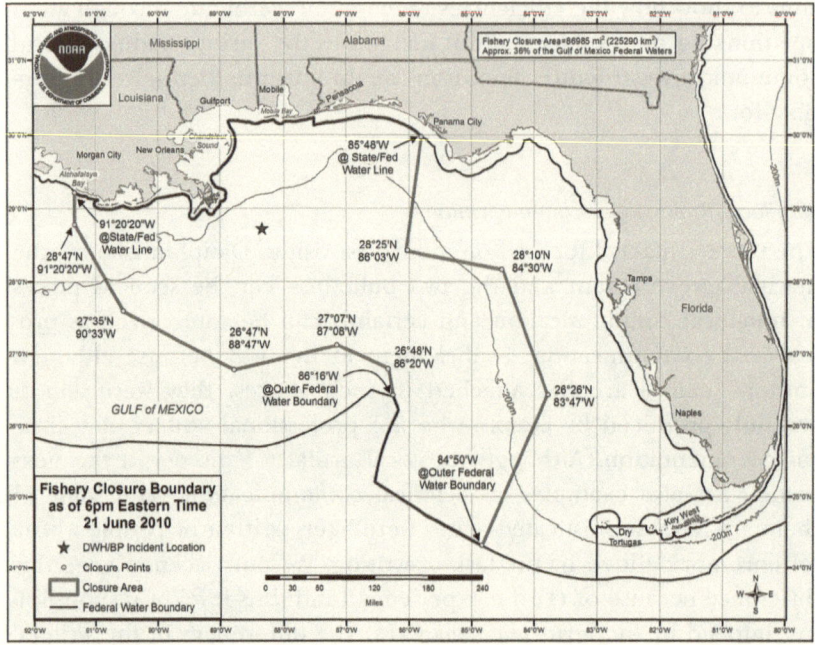

Figure 5.1. Area of federal waters closed to fishing on June 21, 2010, as a result of the Deepwater Horizon spill

a source of considerable economic income. Fisherpersons supply local restaurants with fresh catch every day, and boat captains make their livings giving tours to visitors. Because of these continual, everyday actions—which existed well before any of the current users of the marina were born—certain patterns of life at the marina are deeply ingrained. At the marina in June 2010, BP used a bulletin board to recruit local boat captains to work on oil cleanup. Most of these people were out of work because of government restrictions on fishing as a result of the oil spill, so the offer to hire out themselves and their boats for a different purpose came at a kairotic moment (figure 5.1). The advertisement, then, functioned as an actant seeking to minimize overall local perceptions of risk. It sought to underline BP's attention to social responsibility by explaining how the company was paying for private contractors to aid in the oil cleanup; although those private contractors could no longer book fishing cruises for tourists because of the oil spill, they could still make an income by working for BP. This argument seeks to locate risk in economic concerns to the exclusion of ecological ones; it seeks to shift the balance of an already established cultural pattern. This actant was making a point of recognizing a local need and responding to it.

Local oral rhetorics, though, served as a contrasting actant. Although the marina was nearly empty because every boat was now employed—a change from years past when tourism-focused crews milled about the marina waiting to be hired—the two shopkeepers at the marina store indicated that BP's pay rates were lower than what private contractors typically charged and that the economic benefit would not extend to local restaurant owners who depended on the availability of seafood. Thus, within this particular international-local cultural network, environmental risk was constructed as significant by locals in response to a lack of attention by an international actant. Locals could see more and greater ramifications of the spill than could non-local actors; their perspectives transcended their own lives and looked to their neighbors, across economic micro-cultures (food service, fishing, touring), and extending far into the future in a slow spiral. Locals could see the slowness of the ongoing crisis in a way non-local artifacts did not see or ignored.

This rhetorical trend continued in the informational bulletins found at the Pelican Reef restaurant and the Bienville Chevron station. Authored by the federal government—an actant whose cultural goals are already suspect for many working-class southerners post-Hurricane Katrina (Woods 2010)—these bulletins explained how to apply for payments due to lost wages. Again, these non-local risk communication artifacts sought to frame the risk of the oil spill as economic and therefore controllable; again, local sources orally protested the lack of attention to lasting ecological effects in and of themselves as well as the long-term impact of those effects on the local economy. Whereas local concerns focused on oil's harmful effects on wildlife and on the long-term damage to the seafood and tourism industries, federal concerns focused on immediate actions intended to rebuild the government's tarnished disaster response history in the South. In this class of local oral and federal written cultural artifacts, local discourses valued ecology and federal cultural discourses valued economy. Both threads of discourse operated within a preexisting historical reality. These tensions, which are mediated by continued transcultural exchanges, ultimately produced and are still producing new cultural meanings.

The existence of these two bulletins in conjunction with the literature at the marina demonstrates a correlation between the cultural positionings of international and national rhetorics in this context. The voices evident in these rhetorical artifacts are speaking, in some ways, from a similar cultural position. Rhetors representing federal or national and BP or international cultural positions shared a particular valuing of financial resources as a primary concern; they shared an ideology that requires the construction of a socially responsible or responsive ethos; they shared ideas about effective modes of delivery for their messages. All of these considerations contributed to understandings of integrity within a particular transcultural framework.

Of course, differences also abound in the ways these messages move across cultural boundaries. For example, BP and the US government agreed, in the spill's aftermath, that BP would fund compensation for Gulf Coast residents whose livelihoods suffered because of the oil spill.[2] In the wake of this decision, the US federal government—and not BP—provided information on how to access these funds in the Dauphin Island area. We can infer that the US government placed a higher value on getting citizens access to these monies than did BP, which was—for obvious reasons—more interested in putting out literature that persuaded US citizens to take its money not for free but in exchange for work that had to be done regardless. These artifacts, although they are similar in their ironic use of local cultural centers as spaces for delivery and in their construction of risk as primarily economic, nevertheless have different rhetorical goals. BP's already transnational corporate culture privileged efficiency for the company. The federal government's attention to local culture privileged compensation for its citizens and itself. Neither considered the extended length of the crisis for locals, and neither spared significant resources or attention for human health and the ways, as is often the case in environmental disasters (Haas and Frost 2017; Smith 2015), the poorest communities would be the most affected.

As the slow crisis of the DHD unfolded and non-local actors began to move out of crisis mode, the efficiency models that were privileged became more clearly about appearing—if not actually being—socially responsible. For example, on November 22, 2013, I visited the local Deepwater Horizon Claims Center (3976 B Government Road). Arriving before 1:00 p.m., I discovered a sign out on the door saying they were closed for lunch until 2:30. There was also signage barring cameras and cell phones, including signage that was fairly aggressive about no photos being taken. The fact that the Claims Center was, shall we say, lightly manned a scant three years after the disaster suggests that both the federal government and locals had moved on from the economy model of efficiency under which much of the DHD rhetoric operated. In other words, hegemonic powers were successful, in this case, in locating the prime efficiency as economic—and, thus, more temporary than other efficiencies.

Whereas a feminist perspective might advocate for sustained attention to ecology and health—and, yes, economy as well—the efficiency models in place as the DHD unfolded were focused on economy. Economy is, of course, eminently more recoverable than ecology or human health.

Local and Regional Citizen Communication

What I describe as citizen communication includes those artifacts representing local and regional perspectives. These citizen communicators can all be identified as members of localized cultures of Dauphin Island or the surrounding region by virtue of the fact that their workplaces are on the island. These communicators tend to see more nuance than do non-local communicators because for them the disaster began and continues on a longer time line; they see slow crisis. In addition, these communicators find their entire lives and identities impacted by the DHD rather than just their professional selves, and so their perspectives tend to be intersectional and nuanced. Finally, the label "citizen" implicates these writers in some of the same ways I have tried to implicate myself. For better or worse, they have a significant stake in this particular community—which extends well beyond the stakes of other communicators temporally—and they write from within that particular cultural position.

I have mentioned that the written citizen communications I observed took place exclusively in digital formats. Each of these digital artifacts occupies a unique cultural rhetorical situation and represents different institutional, political, cultural, economic, and ecological forces. Importantly but beyond the scope of this book, these artifacts are also more widely accessible to a broader range of audiences. Despite their nuanced perspectives and differences, the Dauphin Island Real Estate company blog, the Town of Dauphin Island's website, and the Dauphin Island Sea Lab website do share some broad similarities—the foremost of which is their attention to ecological risk as an issue of paramount importance in regard to the Deepwater Horizon Disaster's impact on Dauphin Island. All of these artifacts produced by citizens shared the particular cultural value of attention to ecological risk.

The DISL created the most rhetorical distance between itself and its readers. The DISL, though located on and named after Dauphin Island, is a statewide venture with a focus on research and education. It has a number of allegiances; by the nature of its mission, its most important membership in a cultural group is defined by its valuing of natural resources. At the same time, its public funding places it in a politically precarious situation. During the summer of 2010, the DISL website functioned largely as an actant that legitimized other sources of information. For example, the DISL's oil spill blog identified the Alabama Department of Public Health as having produced "THE BEST review of the toxic issues associated [*sic*] with the oil spill and dispersants" (Dauphin Island Sea Lab 2010). The DISL was largely concerned with the potential ecological consequences of the spill; most of its entries dealt with concerns over oiled wildlife and "sick fish." Although the impact of sick fish on the seafood industry was mentioned, the DISL framed injured wildlife mostly as a problem in and of itself. The DISL engaged in rhetoric about the oil spill mostly in its online space. When I visited the DISL's estuarium and public aquarium, exhibits (which are updated from year to year) did not include discussion of the Deepwater Horizon Disaster. Whether this was a conscious choice or the result of some other issue—funding or time constraints, perhaps—is impossible to know; however, the effect was that the DISL did not compete with federal and international entities in establishing risk-constructing rhetorics on the island. In a network we can recognize as having international, federal, regional, and local actants, this particular regional actant chose not to use a space where it could have established an authoritative voice that would have impacted transcultural understandings of the disaster.

The Town of Dauphin Island's website offered three photographs depicting response to the Deepwater Horizon spill. The first shows oil containment boom stretching into the distance in the waters off Dauphin Island. As much as possible, Dauphin Island surrounded itself with boom in the summer of 2010 in an effort to keep oil off the beaches and out of estuary waters. The second and third photographs show the construction of dunes on the island. The town financed the construction of one dune behind the beaches on the island's gulf side in response to hurricanes in 2009 and 2010; in preparation for the oil spill, this dune was reinforced. More remarkable, though, was the construction of a second dune nearly at the water's edge, intended explicitly to keep oil off the island. These photos are clearly concerned with ecological rather than economic matters; that is, photos were of landscape, not of storefronts, and the website contained no pleas to patronize local businesses. The Town of Dauphin Island was no doubt concerned about economic implications, but this particular cultural artifact constructs risk as ecological.

Although the Town of Dauphin Island, like the DISL, did not explicitly discuss the risk associated with the oil spill in any physical space on the island that I was aware of, it did allow other forms of visible rhetorics that assisted in constructing understandings of risk. BP hired workers to patrol beaches all along the Gulf Coast during the days when oil continued to flow from the Deepwater Horizon well, and Dauphin Island was among the communities that received such aid. During my time at Dauphin Island, BP workers dressed in "hazardous materials" (HAZMAT) suits combed the beach in search of evidence of oil. Although the beaches were clean during my stay, the sight of HAZMAT suits on the beach certainly indicated a risk of incoming oil. We might interpret the presence of these workers and their gear as a manifestation of ecological risk approved by international (BP), federal (US government), and local (the Town of Dauphin Island) actants; these workers were visibly prepared for and protected from incoming oil. The presence of people in bathing suits on the same beaches as these BP employees created a very different message. By their willingness to appear on the beach unprotected, these actants denied the severity of any sort of risk associated with incoming oil. In sum, then, this situation depicts international and federal actants acknowledging the ecological risk of the oil spill. We also have a local actant—the Town of Dauphin Island—joining in this message. At the same time, we have other local actants—that is, beachgoers, visitors, and residents—undermining this acknowledgment of ecological risk. This network, then, shows a tangle of transcultural alliances and loyalties that make the cultural construction of risk fairly complicated.

The DIRE company, meanwhile, operated based on different stakes. Whereas the town government and the DISL were focused on ecological issues, DIRE makes a significant percentage of its income from summer vacation home rentals; the company was therefore far more interested in promoting an understanding of high economic and low ecological risk, which would be the most likely combination to entice visitors not to cancel their summer vacations. DIRE did clearly feel a sense of ethical duty to pay close attention to ecological risk and was scrupulous about updating readers on the presence of any oil; this conscientiousness served to underline the message that ecological risk was, at that time, minimal. In the first entry on its oil spill blog, DIRE directly addressed tourists and visitors, offering to keep them up to date on information and telling them that "no oil has come ashore" and that "it is entirely too soon to automatically assume the worst" (Dauphin Island Real Estate, Inc. 2010). When traces of oil did come ashore on June 1, DIRE reported this fact and began more frequent updates on the amount and size of tarballs, often asserting that the beach was largely clean. DIRE added video updates and technical documents like the fifty-eight-page emergency permit application from the US Army Corps of Engineers to create a new dune fronting the island to protect it from oil (Dauphin Island Real Estate, Inc. 2010). Later, DIRE's blog again addressed tourists, noting that the company had many spaces left for vacation rentals should anyone wish to book a last-minute trip. The implied message, of course, was grave economic risk for local businesses; this stood in opposition to DIRE's construction of the ecological risk as under control. Thus, in this network, DIRE shared a desire to reduce transcultural constructions of ecological risk with the beachgoers mentioned earlier. This desire exists in opposition to the aims of other local and regional actants, who wanted to point out the ecological risk facing their community as a way of underlining the seriousness of the disaster and perhaps as a way of attracting federal and international relief funds.

Local rhetorics about the DHD since 2010 seem largely to have followed the path of non-local/professional rhetorics. That is, in the face of the vast attention paid to economic recovery, concerns about human health and ecology were largely absent. Only the scientists continue to talk about ecology, and the local digital sites discussing human health are gone, except in internet archives. As above, feminist projects would have advocated collaboration and sustained attention to human health and ecology. In thinking about the ways such a response might have been structured, I can't help but return in my mind to the architecture of the Mobile County Health Department's main building, which was designed to separate and divide people into categories. The efficiency of this design is based on the notion that health concerns can be compartmentalized, isolated, treated individually. As any good physician (or person inhabiting a body) can tell you, that's not the case. Like the human body, the DHD contains many interlocking parts, particularly for citizens of the Gulf Coast. The efficiency model the government used to respond to the DHD was based on economic recovery, and local rhetorics advocating other approaches were unable to sustain themselves—at least not in public places.

Guerrilla Media

The notion of guerrilla media (Ding 2009) helpfully complicates the distinctions between local and non-local in relation to Dauphin Island's reactions to the oil spill, and it plays nicely with feminist efficiency models dedicated to mobility, accessibility, permeability and critiques of assumed efficiencies and temporalities. Initially, local groups used digital media to make their voices heard. However, it is not clear whether these outlets could be classified as "guerrilla" because although they were certainly points of tactical intervention, they also operated in temporarily stable, sanctioned locations in the digital world (as opposed to more fleeting media like text messages or chat rooms). Meanwhile, professional actors used more than one method to make their versions of risk heard. For example, not only did BP use globalized means that, for an international company, are more traditional—such as websites and press releases, which I have not discussed because they seemed to have a minimal effect on discourses in the Dauphin Island community—but the company also used what we might think of as a guerrilla campaign to intervene in culturally saturated spaces previously controlled by locals. Operating from a definition of guerrilla media that includes messages that open up new spaces, seek to include marginalized groups, and create alternative political situations, I could argue that the professional artifacts discussed above do, in fact, qualify as guerrilla media. They exist in spaces that are new to them as rhetors and that are valued by cultures operating at local and regional levels; in other words, they open up a transcultural channel of communication that did not previously exist. They seek local and regional audiences who have been most victimized, across intersectional axes, by the disaster. And they attempt to provide useful and productive economic options for those citizens even while recognizing the long-term nature, or slowness, of the crisis.

However, the existence of risk communication artifacts authored by hegemonic forces in culturally important local and regional locations may not, in itself, cause them to be defined as guerrilla media. The situation does lead to some pressing questions. Do we value these artifacts more highly because they demonstrate rhetorical savvy in working transculturally by bringing their messages to locally important spaces? Or do we consider the colonial implications and devalue them because of their infiltration of spaces they should not attempt to claim? In trying to classify these artifacts, we might consider Ding's (2009) portrayal of guerrilla media as allied with Michel de Certeau's (1984) description of tactics users. We know that tactics are weapons of the weak and exist temporally, not geographically. Because the artifacts described above are certainly not weapons of the weak and because they do occupy specific physical locations—whatever those locations may be—we must consider that they actually exist in direct opposition to the principles of guerrilla media. These artifacts, though they are surely examples of purposeful transcultural communication, may actually serve to deaden the potential advantages of transcultural meaning making in this case precisely because of their infiltration of foreign (to them) cultural spaces as well as the one-way nature of their communication. The authors of *Improving Risk Communication* (National Research Council 1989, 151) make recommendations for the process of producing risk communication, including "a spirit of open exchange" that is not limited to technical issues as well as "early and sustained interchange." In creating transcultural technical documentation that constructs risk but does not provide for feedback, these actants diminished the ability of their message to move back and forth across cultural boundaries.

Meanwhile, the artifacts produced by citizens at the regional and local levels also showcase many of the characteristics of guerrilla media. Ding (2009, 330, emphasis added) defined guerrilla media "as interpersonal communication technologies widely used by and easily accessible to the general public, including *technology-assisted media.*" Ding further described guerrilla media as "flexible, mobile, and accessible" (330). Similarly and more recently, Veronica Garrison-Joyner and Elizabeth Caravella (2020) argue for comics—certainly considered an alternative mode in healthcare communication—to help address deficits in health literacy and the resulting negative health outcomes. While I did not find Dauphin Island citizens using comics specifically, Garrison-Joyner and Caravella's work highlights that anyone who is communicating healthcare needs to think more broadly about accessibility. It is important that Gulf Coast citizens took up digital spaces—to the apparent exclusion of non-digital spaces—in their quest to narrate for themselves the risk facing their communities and their cultural values. They found these digital spaces to be more flexible, mobile, and accessible for their purposes. In taking up digital guerrilla media, the citizens of Dauphin Island not only intervened tactically in the construction of risk but they also intervened in the public, transcultural construction of voices that were allowed to narrate the disaster. Indeed, the DIRE oil blog (Dauphin Island Real Estate, Inc. 2010) directly addressed the need for transcultural communication from guerrilla sources in its first post:

> Again, not a sign of any oil so far, and we certainly hope that the fragile environment of Dauphin Island can avoid being impacted. We cautiously have our fingers crossed! Many of the national media reports over the last several days have implied that not only Dauphin Island, but Gulf Shores/Orange Beach and the Florida Panhandle . . . even locations as far removed as Key West, are basically going to be devastated by this oil spill. It is certainly true that the spill is an event of national significance, but we feel that it is entirely too soon to automatically assume the worst.

In this passage, the DIRE blogger(s) hinted that mainstream media sources—which are typically able to send messages across, within, and between cultural groups with ease—are not reporting accurately on local conditions. As such, DIRE purposefully sought a space in which to voice transcultural messages—messages aimed at the many tourists and repeat visitors from across the nation and the world, representing countless cultural groups and intersectional identities—that reconstruct in more productive ways the risks posed to their community.

As guerilla media is wont to do, many of the messages I would classify in this way have either disappeared or become exceedingly difficult to find. Because the communicative acts I classify as guerrilla in this case tended to be those that were local and digital, they quickly became the victims of search algorithms designed to privilege what's new, what's sponsored, and what gets a lot of traffic. They also faded away as domain names expired, the economy recovered, and author-critic-journalists lost interest. In short, precisely because they aimed to respond to feminist efficiency models that valued what is fast, responsive, and collaborative, they also were ephemeral.

(Feminist) Transcultural Communication

Feminist analysis of a complex communicative situation like the DHD not only invites but requires a transcultural approach to help make sense of rhetors' different obligations and interests and to do at least some justice to the intersectional interests involved. I have used efficiency as a way to think about the values that underly those obligations and interests. Transcultural analysis centers on an understanding of a "new global cultural economy" that "has to be seen as a complex, overlapping, disjunctive order that cannot any longer be understood in terms of existing center-periphery models" (Appadurai 1996, 32). In other words, transcultural analysis—like feminist inquiry—relies on understanding that no cultural artifact exists in a vacuum. Rather, each of the technical communication artifacts discussed above looks the way it does because of its dependence on other artifacts that are also caught up in transcultural flows of information and that constitute significant parts of deeply engrained rhetorical ecologies. (And each of these artifacts is the result of its writer's intersectional identities.) None of these artifacts can be identified as the central source from which information flows. Thus, transcultural analysis—like intersectional feminisms—calls into question traditional understandings of the center and the margin. The communications of massive government and corporate entities—which may largely write the event for history textbooks—are not the center because they omit important experiential data, but neither are the rhetorics produced in Gulf Coast communities the central voice of this disaster. These sets of communications are interdependent and always responsive to each other, even as they develop contradictory narratives.

Further, transcultural analysis helps maintain important critiques of the production of master narratives. From the perspective of a few years into the future, we can easily buy into a master narrative that privileges focus on the long-term environmental consequences of the Deepwater Horizon Disaster. As an example, I found an account of the disaster in a rhetoric textbook from which I teach. Sharon Crowley and Deborah Hawhee (2012, 110) wrote that "before the well could be capped, fisheries were endangered and beaches were polluted. Everyone concerned regarded this event as an unmitigated disaster." Crowley and Hawhee's account privileges what they expect will matter to their broad audience of contemporary students, which is the long-term environmental effect on ecology and American lifestyles. However, they do not mention the long-term economic impact on small Gulf Coast communities—the slow crisis that we know this actually is—a perspective that becomes apparent only when we pay attention to transcultural flows of information, even though, as we have seen, it is wildly apparent in other contexts. Without a transcultural approach, we risk losing perspectives that are only evident when we take into account the multiple networks and communicators with stakes in the situation. If we remove any one of the international, national, regional, or local perspectives discussed in this chapter, we risk flattening or losing altogether voices and interactions that are important parts of a set of co-constructed understandings of risk. In other words, doing this sort of analysis without including regional and local communicators' impact on the transcultural flow of information risks covering over the voices of the very people the aforementioned master narrative seeks to protect.

Ultimately, transcultural analysis done with attention to an interdisciplinary methodology can help resituate power relations between risk communicators who write from different institutional, geographic, organizational, and cultural locations. Transcultural inquiry "alters our relationship with others. It asks us to acknowledge the expertise and agency of people" not traditionally considered experts (Flower 2002, 197). Paula Chakravartty and Yuezhi Zhao's (2008, 10) "transcultural political economy" approach highlights the ability of transcultural work to "integrate institutional and cultural analyses" and to make apparent sites and systems of power. In other words, transcultural analysis asks us to reconsider the space in which cultural knowledge making occurs.

I have already begun to draw on literal and metaphorical understandings of space or location. Many technical communication scholars have already pointed out the importance of the physical space that surrounds users as they interact with technologies (Barnett 2012; Dolmage 2014; Hurley 2018; Kimme Hea et al. 2012; Welch 2005). M. Jimmie Killingsworth (2005, 360) emphasized that writing always "comes from somewhere and bears the marks of its place of origin even in a mobile, globalized, networked community of discourse users." Further, Huatong Sun (2009) studied the localized nature of mobile technologies while also emphasizing the importance of the terms we use to describe the physical locations of technologies and communication. Finally, space is an important consideration because it leads to discussions of colonization and ownership. My surprise at the placement of particular communication artifacts—non-local actors using bulletin boards, for example—arises from my awareness of already established claims to particular spaces. For example, the fact that I consider "local" contexts as places where a transnational corporation like BP is not native assumes that ownership of that space belongs to people I classify as local; this assumption is not arbitrary but rather is rooted in culturally constructed beliefs about ownership of space.

Cultural theorist de Certeau (1984) began a trend in which scholars in a variety of related disciplines, many of whom interact with technical communication, have taken up inquiry into the politics of space. Roxanne Mountford (2001, 41, 42) argued that a rhetorical space is "the geography of a communicative event" and that these spaces "carry the residue of history upon them." Jenny Edbauer (2005, 9, original emphases) found "a connection between certain models of rhetorical situation and a sense of *place*" and also theorized a "framework of *affective ecologies* that recontextualizes rhetorics" temporally, historically, and spatially. Richard Marback (2004, 25), drawing on Mountford, advocated the creation or usage of rhetorical places "which generate in us commitments and which demand from us respect." Nedra Reynolds (1998, 13) theorized a "geographic turn" that calls us to "interpret . . . spatial metaphors as 'imagined geographies' " that define identities and to argue for attention to a "spatial politics" in composing that seeks to make invisible spaces more visible. W. Michele Simmons and Jeffrey T. Grabill (2007) discussed what actions must be taken to make possible particular kinds of communication in particular kinds of places, and Renee M. Moreno (2002, 233) argued that "there are locations dominant society cannot touch," therein framing the ownership narratives I mention above. All these articulations of the importance of geography to communication come from theorists in the intertwined fields of rhetoric and composition—fields that are themselves intricately tied to technical communication—and they all explicitly tie geography and place to specifics of culture.

This connection is important because technical communication artifacts are always culturally saturated, loaded, and implicated. The situation I focus on requires transcultural critique because authors and their intersectional identities are often not apparent, but they have immense rhetorical power. Further, each of the artifacts I discuss exists in a complex place both rhetorically and physically. I deal with transcultural agents ranging from BP—an international corporation with an interest in establishing particular beliefs in communities outside its country of origin—to small-town private citizens, who also do transcultural work by reaching into digital spaces to come to voice. Linda Flower (2002, 186, original emphasis) worked to define community inquiry—a process citizens of Dauphin Island certainly engaged in during the oil spill—as an intercultural process wherein "the partners in an intercultural inquiry attempt to *use* the differences of race, class, culture, or discourse that are available to them to understand shared questions." Transcultural inquiry, by this definition, is about reaching across a lack of shared experience to discover "rival readings of that issue that have the potential to transform both the inquirers and their interpretations of problematic issues in the world" (186).

This definition of transcultural inquiry, then, also provides a working definition of culture: any community created by connections in identity, discourse, interest, or values. Culture is the sort of term that can easily become a reductive monolith. Many scholars, in particular some who study Mary Louise Pratt's (1999) contact zones, have warned against monolithic and binary constructions of culture (van Slyck 1997). The work I do here—and any scholarly work—will necessarily deal in monoliths at times. We have to; it is necessary to hold a thing—a communicative act, an artifact, a rhetorical situation, a perspective on feminism, a crisis, a disaster—still for a moment to understand its complexity.

TOWARD NEW EFFICIENCIES

Digital guerrilla media allowed citizens of small communities like Dauphin Island to come to voice in a way that was not possible a few decades ago. However, as mentioned, the structures of internet search engines privilege already established content (Potts 2014). Even as local communicators recognized what I have termed *slow crisis*, their writings about the situation were always already ephemeral. Most of the sites that come up today when one searches for Deepwater Horizon information online are authored by BP or major news networks. Charles Kostelnick (2007, 21) wrote that risk communication is a situation "in which the stakes are high for participants in the communication process and in which something almost always can go wrong." For businesses and citizens on Dauphin Island, the transcultural privileging of rhetorics of risk other than their own could certainly constitute something going wrong. I saw the effects of what went wrong during my stay in June 2010, when in one glance I could take in pristine beaches and myriad precautions against oil as well as closed restaurants and empty rental properties. Despite attempts by local communicators to construct transcultural understandings of risk in ways that would be sustainable to the community, those voices were subsumed under more powerful rhetorics of risk. This situation provides a case study underlining the need for technical communicators to be aware of the cultural construction of risk and a resulting ethical obligation to listen to more diverse (intersectional, transcultural, long-term–oriented) voices in order to see risk as localized and contextual.

The purposes of risk communication, according to the National Research Council (1989), are twofold: to inform and to influence. Successful risk communication "makes for better-informed decision makers" (80). The artifacts surveyed here were certainly designed to inform and influence so that readers would make particular decisions. However, all of these artifacts placed a high value on adhering to a particular cultural understanding of objective reporting. They made rhetorical moves to maintain rhetorical distance and to prove that they were not going to "project the contents of their own heads" onto the supposedly external and neutral existence of a risk situation (Belenky, Clinchy, Goldberger, and Tarule 1986, 109). James F. Stratman (2007, 25) suggested that "if this notion that persuasion is inherently unethical remains a dominant view in the risk communication field, then rhetorical theory will most likely not find a welcoming audience there," a possibility he frames as unfortunate. Other scholars have also noted that technical communication, including risk communication, cannot "exist in a vacuum, separate from larger social and cultural relations" (Thralls and Blyler 2002, 190). Rather, documentation that we deem to be technical or objective operates in such a way as to keep us from seeing the ways it supports particular cultural values. Therefore, we need to expand our definition of risk to include culturally informed citizen constructions and assessment of risk; we need to pay attention to nontraditional modes of delivery and trace out the rhetorical and cultural effects of such reversals. Doing so can only help to contextualize transcultural risk communication practices like those that surrounded the Deepwater Horizon Disaster.

This study shows how transcultural risk communication messages made meaning for specific local and global cultural groups at a specific time and place. It emphasizes the importance of transcultural messages that slip and slide from local to global spaces and back again, and it demonstrates the rhetorical importance of the delivery locations of risk communication artifacts. Studying risk communication in both time- and place-sensitive ways can help us see the impact of risk communication on transcultural audiences, and it provides information about how a variety of actants think of technical communicators' ethical responsibilities. I hope that this, in turn, can help technical communicators think about how to manage those responsibilities in terms of discourses surrounding the aftereffects of disasters.

Further, I underscore the resilience—a well-explored feminist value (Flynn, Sotirin, and Brady 2012)—of many of the cultural groups in question. First and foremost is the survival of the delicate ecosystem—including human presence—on Dauphin Island. Although many sources acknowledge that Gulf Coast communities face "years of clean-up" (Dauphin Island Sea Lab 2010), it is also true that local fisherman are back at work (Thierry 2011) and most of Dauphin Island's tourism-dependent businesses survived the disaster. Even as local actants recognize slow crisis, they also create ways to interrupt and live within it. By actively constructing cultural meaning in online spaces, Dauphin Island citizens tactically intervened in the ways outside forces were depicting the oil spill and its impact on their community. In their book-length work on globally networked learning environments, Doreen Starke-Meyerring and Melanie Wilson (2008, 6) said "digital network technologies thus allow individuals and organizations to reach out to new networks, to build new relationships, and to reach for new opportunities for growth"—all of which are transcultural means of meaning making and community building. The ability of Gulf Coast residents to survive and even thrive in the wake of the Deepwater Horizon Disaster is due at least in part to the ways community members are able to understand the nature of slow crisis while also mediating transcultural understandings of the risks they face when such disasters arise.

6
LOOKING FORWARD, LOOKING BACK

HISTORIES

The Deepwater Horizon Disaster (DHD) attained widespread public knowledge in April 2010, when an explosion onboard an oil rig in the Gulf of Mexico killed eleven men and triggered a months-long oil leak. BP PLC has taken most of the blame in the public eye, although subcontractors including Transocean have also been held legally liable. The legal and environmental ramifications of this disaster continue to unfold, complicated by tension between a foreign oil company and domestic courts. The primary focus of all parties seems to be on economic effects, with a sub-focus on environmental effects and little to no discussion of effects on human health. This focus has deeply impacted how the history of the DHD has been constructed in popular media.

The Deepwater Horizon Disaster really began in 2008—if not earlier. That year is when the Minerals Management Service (MMS, a division of the US federal government) agreed to lease mineral rights in the Macondo area to BP. The Macondo area, a region forty miles off the coast of Louisiana, was promising enough for BP to send its Deepwater Horizon oil drilling rig (which BP was leasing from Transocean Ltd.), one of a relatively few rigs that could operate in more than 5,000 feet of water. The Deepwater Horizon rig had previously completed the deepest oil drilling endeavor in history (more than 35,000 feet), although little was made of this in the popular press at the time. That changed in April 2010, when the explosion onboard the Deepwater Horizon rig at Macondo precipitated the largest marine oil spill and the fifth-largest oil spill in the history of the world. Indeed, the history of the DHD evidences many extremes—extremes of profit, extremes of risk, and, most important, extremes of efficiency.

The Deepwater Horizon Disaster really began in 1908—if not earlier. That year is when British explorers found oil in Iran, paving the way for the formation of the Anglo-Persian Oil Company (APOC). APOC continued to drill in the Middle East while expanding into new markets. In 1954, after a series of name changes, it became the British

Petroleum Company. By 2001, owing to a dismal safety and environmental record, the company—now one of the largest oil companies in the world—rebranded itself as BP and rolled out a green sunburst logo. Not even five years later, more than 170 workers were injured and 15 were killed in the Texas City Refinery explosion.

The Deepwater Horizon Disaster really began in 1699—if not earlier. That year is when the Chitimacha Tribe of Louisiana encountered European explorers who began kidnapping and enslaving tribal members. The Chitimacha eventually retaliated, and the ensuing war lasted twelve years until a peace treaty was signed in 1718. This treaty was the first in a series of legal documents used by Euro-Americans to reduce tribal lands from more than 5,400 acres to 262 acres. The tribe now owns 963 acres, a small portion of its original territory in southern Louisiana, and does not control water rights off the coast (Sovereign Nation of the Chitimacha 2015). Thus, the US government, through the MMS, was able to commission surveys of the area in 1998 and 2003, making it possible to lease mineral rights (which allow the owner of said rights to mine minerals below the surface of the property) for Mississippi Canyon Block 252 (codenamed the Macondo area) to BP in 2008.[1]

These different available histories call into relief the ways different notions of efficiency—of what matters and why—create different stories. These histories also highlight areas where certain bodies aren't apparent. So often, what we're **not** talking about is bodies; we talk about land rights, economic development, ecological harms—but major histories of this event leave out people.

SOME NOTABLE PEOPLE

At one time, I set out to examine the effects of oil spills on human health. I was not particularly interested in oil but rather in the larger effects of environmental health on human health. I wanted to know more about the connections between the environment and our bodies and why we talk so little about what must be a massive effect. That is not the project I ended up writing, but it doesn't mean that others haven't been at work on it. For example, Danielle Koonce (2021) says we *have* research on health impacts for a variety of environmental factors; what we *need* is research on organizing. Koonce researches racialized environmental impacts, specifically related to the location of hog farming operations, and she has recognized this problem and points to action-oriented approaches to it.

In my quest to make bodies and people more apparent, I think it's worth pausing to spend some time with the people—the women—who have been doing the work of dealing with human health as it relates to environmental crisis. When I realized that the DHD was generating so little interest in human health effects and that one of the only early examples of this interest had to do with pregnant women, I was initially frustrated. In attempting to figure out why the only urgency applied to fetal health, I returned to Indigenous activist Katsi Cook (2004, 158) for context:

> Science tells us that our nursing infants are at the top of the food chain. Industrial chemicals like PCBs, DDT and HCBs dumped into the waters and soil move up through the food chain, through plants, fish, wildlife, and into the bodies of human beings who eat them. These contaminants resist being broken down by the body, which stores them in our fat cells. The only known way to excrete large amounts of them is through pregnancy, where they cross the placenta, and during lactation, where they are moved out of storage in our fat cells and show up in our breast milk. In this way, each succeeding generation inherits a body burden of toxic contaminants from their mothers. In this way, we, as women, are the landfill.

Returning to efficiency, two points become important. First, the timing of the CDC's release—weeks after the Deepwater Horizon Disaster—suggests that governmental concern may have been less about getting urgent news out to pregnant women than about demonstrating that it had made some attempt at social responsibility. Second, the government's response being focused entirely on pregnant women reinforces a troubling trend critiqued in a great deal of feminist literature—that of women as fetal containers. The CDC's release is, after all, aimed at the health not of women but of fetuses. This, then, is an example of documentation that appears to adhere to god terms that are familiar to technical communicators—objective, neutral, efficient—and that also appears well-intentioned. However, a further examination of the full context of the release suggests that its claims to any of those adjectives may be suspect.

Such suspicions are not news to the people who have been doing the work of environmental justice—the people whose bodies and communities have been on the line all along. As an example, "Led primarily by working-class women and women of color, the environmental justice movement has its roots in the poor, predominantly African American community of Warren County, North Carolina" (Murphy 2017, 157). Mollie K. Murphy differentiates between environmental justice and the "antitoxics movement," which we might think of as a larger umbrella term for a movement whose members are concerned with environmental health but not necessarily aware of or interested in the unequal effects of environmental toxins and poor communities and communities of color. Interestingly, the antitoxics movement, too, is largely women-led. As numerous scholars have hypothesized (Cook 2004; Cuomo 1998; LaDuke 1999; Murphy 2017; Unger 2010), this is likely because women are the first environment—the people who bear the brunt of reproductive effects of environmental toxins.

The connection to maternity is an important and ubiquitous element of the framing of environmental work: "Environmental justice and antitoxics activists both appeal to motherhood, a powerful but highly contested trope in social movement activism. Maternal appeals have the power to legitimize experiential knowledge and bring seemingly irrelevant, 'private' matters to public debate" (Murphy 2017, 158). We see this in the case of Wilma Alpha Subra—president of the Subra Company, winner of a MacArthur grant, and a chemist and environmentalist. Subra, who is based in southern Louisiana, works to communicate with the public about toxic chemicals that may affect them (MacArthur Foundation 2005) and to advocate for local and Native communities. Media coverage of Subra's work tends to label her an "activist grandmother," an interesting combination of terms that draws attention to both her femininity and her matronliness and seems, at times, intended to serve as a juxtaposition to her toughness.

The daughter of a chemist and the granddaughter of a fisherman, Subra knows that understanding the science of a problem related to environmental and human health is critical: "Looking at all the environmental issues, you had to understand what the impact meant, and put it in terms the government agencies could respond to" (cited in Goldenberg 2010). With a master's degree in microbiology and chemistry from the University of Louisiana at Lafayette,[2] Subra is as much a community organizer as she is a scientist: "I'd go out a lot at night and help communities organize, help them understand the technical issues, and then take samples for them to test their water or soil" (cited in Mullin 2020, para. 13). When interviewed in June 2010 by the *Guardian*—after testifying before a congressional committee investigating the DHD—Subra said she had so far received 300 to 400 complaints related to the DHD (Goldenberg 2010). "Headaches, dizziness, stinging eyes, some chest pains," she told the newspaper. "They come in at night very sick, but they need that job, so they go out again the next morning" (para. 4). Subra worked proactively—an unusual approach in disaster timelines and one that certainly complements feminist efficiencies of cooperation and care—with the Louisiana Environmental Action Network to try to get protective gear to cleanup workers.

Importantly, Subra's persistence—her apparency, collaborative work, and privileging of apparently feminist efficiencies—has made her a target. The following passage from a 2020 *Chemical and Engineering News* report is telling:

> "I've had break-ins," she says. "One day when my glass man came to change the glass, he said, 'This is the last time. Call the guy in town and get him to put up burglar screens.' So now I just get rotten eggs and things like that."
>
> And then there was the drive-by shooting several years ago. "My husband was working in the flower bed about 7:00 p.m.—here you wait till things get a bit cool," she recounts. Their home is around the corner from the office. "This car was going by really slow, and when it came this way, the passenger shot at the building."
>
> When the police stopped the car at the traffic light up the road, Subra says, the gun and passenger were gone. "They wouldn't tell me who the driver was because it would infringe on his civil rights," she adds.
>
> Subra assumes that local people are being paid to harass her. "I can always name two or three people who are opposed to what I'm doing," she says. No doubt: Subra has spent decades challenging industry on development and environmental issues and educating communities near industrial operations to do the same. (Mullin 2020, para. 7–10)

For Subra, this is a day at the office. Apparency, then, is a dangerous business. Feminist apparency in particular can be a game of high stakes. And yet, without making bodies more apparent, we are left without a way to protect ourselves.

Raquel M. Robvais (2020) points to the dual nature of apparency in her witnessing of sickle cell disease (SCD) sufferers'—"sicklers"—impossible situation, as the medical establishment cannot seem to avoid racist preconceptions of Black bodies seeking treatment. SCD, a genetic condition that disproportionately affects those of African heritage, can cause many complications, including chronic pain: "Those suffering with pain often have to engage in a performance experience that entails having the correct attire, knowing the appropriate jargon and knowing the correct amount of medicine and so forth, in order to receive respectable treatment" (8). For SCD patients, embodied apparency is an important part of receiving care. Indeed, an SCD patient's approach to apparency can determine *whether* they receive care or are instead treated as a "drug seeker." Robvais continues: "The emergency room for a sickle cell patient is a place where blackness functions as a historical albatross and a liberating freedom" (8). That liberation comes in the form of community. Many SCD patients, Robvais shows, learn how to interface with the medical establishment by tapping into scripts that benefit them: "What is evident in the lives of SCA patients are transgressive acts cloaked as seemingly mundane, pedestrian acts like dressing a particular way before going to the E.R., like going to the E.R. on a particular day, making sure you don't visit the E.R. alone and drinking plenty of water prior to the visit, just to name a few" (10). Like apparent feminist theory, our bodies are subject to others' interpretations. They are permeable. We are vulnerable in our resilience, and this makes apparency a complicated venture.

THE APPARENCY OF BODIES

In reimagining origin stories, as I do at the beginning of this chapter, apparent feminism demands that we consider the agent(s) of such stories. Who gets to tell origin stories, and how are those storytellers embodied? In this section, then, I explore how the apparency of our bodies connects to our ability to produce histories, technical compositions, and technical communication. Barbara A. Biesecker (1989) notes that invoking the term *audience* signifies the presence of a body or bodies. In addition, theorists of rhetoric and technical communication have often considered the effects of a rhetor's embodiment. But what is one to do when one's own body qualifies not as an audience or a rhetor but instead as a constraint? Which bodies are recognized as able to create origin stories, histories, and theories? And what do particular bodies have to do to be heard and to be counted among those producers?

We might think of this history making as a process of technical composition. Robert McRuer (2006, 152, emphasis added) states that "contemporary composition is a highly monitored cultural practice, and those doing the monitoring (on some level, all of us involved) are intent on producing order and *efficiency* where there was none and, ultimately, on forgetting the messy composing process and the composing bodies that experience it." The creation of histories mirrors this process; it is a microcosm of the composition process McRuer describes. To produce coherent histories, technical communicators may feel pressure to cover up "messy" narratives, ideas, and people. It is up to us to redefine efficiency in a way that is explicitly aware of "the messy composing process" and that takes a far broader view of which "composing bodies" are valued.

Apparent feminists should be especially interested in and critical of the ways female bodies are represented in visual culture; technical visible rhetorics are often complicit in producing and reproducing antifeminist representations. As I have maintained, apparency is not the same thing as visuality; however, the two concepts are closely related, and apparent feminists have an obligation to be aware of the visuality of our bodies and of visual rhetorical commonplaces sited on our bodies. In many cases, this will mean intervening in particular genres, as when Kirstin Cronn-Mills (2000) documents a shift to a (relatively) more feminist perspective in marketing strategies by a women's clothing company and when young feminists wear the now popular "this is what a feminist looks like" apparel.

Apparent feminist interventions into practices of visualization may also mean finding more permanent ways to produce an apparent feminist body. But what might an apparent feminist body look like? Since apparent feminism is, like other beliefs and perspectives, not initially visible on our bodies, it is the responsibility of apparent feminists to find ways of making bodies apparently feminist. The presence of female bodies in new places is important; however, presence does not suggest explicitness in the same way apparency does. Producing an apparent feminist body means more than showing up. It means making one's feminism apparent not only to others' eyes but also to their thoughts. It means being recognized as a person—a feminist—with a valid perspective; it means mattering to the conversation at hand. Finding ways to make this happen is a constant problem apparent feminism calls into being.

One way of exploring how to make feminist bodies apparent in technical communication studies is by looking to research on embodiment in digital spaces. Recent scholarship has provided some examples of how to approach this connection. Modupe Yusuf and Veena Namboodri Schioppa (2022) offer a detailed analysis of Black hair tutorials on YouTube as examples of procedural discourses and the ways those discourses shape our bodies. They point out the importance of digital spaces as connections to community when geography introduces challenges—Yusuf having found that Houghton, Michigan, does not have a Black-owned hair salon—and they argue that students and instructors alike can benefit from exploring how technical and professional communication "is being produced in extrainstitutional contexts, identify whether there's a social justice issue that can be addressed, and [describe] how students can contribute to such projects" (279).

Of course, social justice issues to be addressed abound at the intersection of embodiment and digital spaces. Antonio Byrd (2019) positions critical race theory as integral to studies of the materiality of literacy, given literacy's nature as socially constructed and tied to false myths about hard work and perseverance yielding economic comfort and freedom. Adam J. Banks's (2006) germinal work on technology and race oriented the field to issues of meaningful access in tension with practices of exploitation. Safiya Noble (2018) shows how algorithms and other methods of discoverability perpetuate racism and sexism. Charlton D. McIlwain (2019), the 2022 Computers and Writing keynote speaker, both uncovers how technologies are used to police Black Americans and recovers the histories of early Black computer engineers. Manuel Castells (2012) presents polarizing perspectives on how digital worlds affect and are affected by social movements, moving from the Arab Spring to Occupy Wall Street to show how digital world making is fundamentally and entirely a practice rooted in rhetorical meaning-making practices. And cyberfeminists have long shown that the internet is not the democratizing medium some would have us believe (Fernandez, Wilding, and Wright 2003). In fact, the internet represents a wicked problem for embodiment studies and visual rhetorics scholars.

Shannon Butts and Madison Jones's (2021) work on EcoTour, a multimedia environmental advocacy project in a state park, demonstrates one way of helping us better visualize and comprehend wicked problems and their effects on local communities. For another set of examples, in a special issue of *Community Literacy Journal*:

- Several authors engage the notion of the digital divide as it exists in relation to gender, with Melody Bowdon and Russell Carpenter (2011) noting that "digital divides may be closing in some ways, but in many ways they're just moving from one place to another" (3).
- Douglas Walls (2011, 99) writes a largely positive review of Virginia Eubanks's book *Digital Dead End: Fighting for Social Justice in the Information Age* (2011) but also notes that Eubanks does not draw much on rhetoric and composition scholarship: "For example, Eubanks makes claims about the regrettable lack of scholarship on the relationship between citizenship and technology, something that has been considered frequently in rhetoric and composition." This underlines the importance—the *efficiency* to be had through inclusivity—of interdisciplinary work for those interested in social justice.
- David Dadurka and Stacey Pigg (2011, 10) point out the importance of convincing audiences that "race, gender, class, sexuality, and power dynamics do not disappear in web spaces."

I take up Dadurka and Pigg's recommendations for further research (16–17) and utilize an apparent feminist perspective to rearticulate them: technical communication scholars should search out broader understandings of social media usage that include how those embodied as women and those who identify as feminists build online relationships. Further, I suggest that far more work remains in studying who consumes and who produces digital work and how gender affects such patterns of use and consumption. Apparent feminist technical communicators must be active in researching socially just practices for distance education and web design.

Research on women's internet usage—which necessarily invokes the apparency of female bodies—has sponsored some shifts in the way digital research is conducted. Apparent feminist approaches to this research now articulate that women's practices of knowledge making have affected research methods. The result has been a shift toward qualitative research in this branch of technical communication scholarship. As evidence, Cindy Royal (2005, 421) found that "a good amount of the research being done is with qualitative methods that emphasize the cultural complexity of the issues that are difficult to explain purely by the numbers." Royal identifies a number of gendered themes in digital research, including "equal access but unequal usage, visions of utopia and dystopia in technology, and the integration of real world and online worlds" (421). These are subjects that require the application of apparent feminist research methodologies. Apparent feminist approaches to internet research can suggest new models for making apparent feminist and female bodies in ways that effect user advocacy, user-centered design of online spaces, and gendered dynamics in digital rhetorics. In addition, apparent feminist approaches to online research can alter the shape of technical communication curricula in both online and face-to-face programs by making the needs and strengths of female and feminist bodies explicit.

A potentially rich area of focus for apparent feminist technical communicators engaging in internet research might be the rhetorical situation created by anonymous online discussion threads. Conversation threads that follow news stories are an especially intriguing context for study because of their accessibility and their usage by a broad segment of the internet-using population. In my job as news editor for a small-town newspaper in 2009, I found myself—rather unwillingly—in the role of comments moderator. Although I was purposely light-handed in deleting comments, I did note that most of the comments I considered for deletion for violating the rules against personal threats were attacks on users who were apparently female. As a prominent member of the news staff, I was a frequent target. I was never physically attacked, but I noticed that my male colleagues ensured that I never walked to my car alone after work.

Since leaving the newspaper, I have participated in online message boards and have found that my own identification as a woman and a feminist in digital spaces makes my (perceived) physical body a site for others' aggression. In one particularly unsettling situation, I identified myself as a feminist in a comment on a CNN story about the reporting of rape on college campuses. The story framed rape as a crime perpetrated only against women and portrayed only women as capable of being victims. My comment questioned this framing and whether the surveys reported on had included information about the sex, gender, and sexual orientation of perpetrators, victims, or both and the effects of those criteria on tendencies to report rapes. Seconds later, another user posted a response that I won't repeat verbatim; this comment alleged that my self-identification as a feminist makes it acceptable for me to be sexually victimized. While a number of other users jumped in quickly to defend me and the offending comment was flagged and eventually removed, it seems clear that my feminist apparency made me a victim of rhetorical abuse in this situation.

Digital rhetorical spaces, I argue, provide important—but not safe—venues for apparent feminist work. Such spaces allow users to construct apparent feminist bodies that can draw on and benefit from physical embodiment. They also provide a measure of physical—if not intellectual or emotional—insulation from oppressors. In the situation I described above, my feminist apparency and the effects it generated helped me learn just how dangerous feminist apparency can be. This digital attack reinforced my understanding of the risk associated with self-identification as a feminist. This risk of danger or violence is a constant problem that apparent feminists—including apparent feminist students—must negotiate; indeed, it is a continuation of a problem feminists have long faced. In fact, risk is something feminists—and all marginalized people—must take into account in every context, and this will be true until feminism becomes irrelevant. Apparent feminists must perpetually weigh the risks to themselves against the risks of not making feminism apparent and determine the most *efficient* approach in a given situation. For apparent feminist students, navigating the digital rhetorical space of the online classroom is particularly contentious since online interactions will certainly have social consequences; digital pedagogy requires training (Robinson et al. 2019), including diversity training.

While it is important to recognize that digital spaces do not constitute a panacea for situations in which one's embodiment qualifies as a constraint, they do provide new possibilities for apparent feminist technical communicators to intervene in support of social justice. Existing technical communication research has shown some affordances of digital spaces in providing supportive environments for female-identified people navigating health concerns in particular; the same research addresses drawbacks and limitations as well (Haas 2008b; Roundtree 2017; Wray and Verzosa Hurley 2016).[3] Likewise, apparent feminism does not provide easy answers to the questions I posed about which bodies are recognized as producers. However, supporting the apparency of feminist bodies can help us develop new ways of understanding the situated nature of canonical histories. Further, apparent feminist digital spaces can serve as usability testing grounds for articulating feminist apparency in a variety of other—both digital and non-digital—rhetorical situations.

THEORIZING A NEW EFFICIENCY

This book has explored communication about continuing health effects related to the oil spill. I use an apparent feminist theoretical lens—with attending ideas about the place of technical communication scholarship, decolonial practice, and medical rhetorics—to demonstrate how reconsidering understandings of efficiency in such situations subsequently changes possible approaches to acceptable risk related to healthcare. I focus on healthcare communication with the understanding that it functions, in this instance at least, as a repository and reflection of public understandings of risk and health. While researchers in technical communication have addressed the confluence of healthcare communication, risk, and ethics (Ding 2009, 2012, 2013; Lundgren 1994; Youngblood 2012), this particular study is unusual in that it considers environmental disasters as a catalyst for health risks and also questions the role of efficiency in identifying those health risks and producing healthcare communication about them. Crisis communication typically deals only with urgent threats to life; this approach uses rearticulations of efficiency framing to widen what counts as urgent.

Efficiency is often thought of as the balancing point between least effort expended and most result gained. Increasingly, though, corporate and government culture in the United States focuses on the former part of that equation. This is no doubt in part because of the risk involved; greater results justify additional effort, but committing to such effort is a risk because those greater results are not a foregone conclusion. While the hazard here is certainly real, consistent attention to "least effort expended" establishes a pattern and a precedent which mean that the potential good results in a situation are too often rendered unapparent and the potential risk involved thus elided altogether.

The apparent feminist methodology this book takes up focuses on the latter portion of the efficiency equation. In other words, I encourage actors in risk situations to carefully consider the potential result gained, particularly as it affects human bodies. I also argue for attention to which human bodies are named as part of a risk equation, and I insist on attention to reciprocity and justice as part of that understanding. This particular reframing is something that is easy to theorize and harder to practice. Our country's history of social justice activism regarding the ballot shows this rather succinctly. Beverly Guy-Sheftall's searing introduction to her 1995 book *Words of Fire: An Anthology of African-American Feminist Thought* chronicles a history of struggle that frequently includes activists' explicit discussions of whether and why to forward the interests of (white) women or Black people (men) first. Early on, she writes, abolitionists and feminists strengthened one another's positions. But fissures emerged in 1869, when members of the Equal Rights Association began to debate whether linking the women's and Black suffrage movements would "seriously reduce the chances of securing the ballot" (5) and, if so, how they should proceed. Unsurprisingly, a split emerged.

While both groups have now gained suffrage, this struggle over which bodies should be made apparent for activist efforts to be most efficient has not disappeared. Kimberly C. Harper (2021) chronicles a similar conflict in 2020, when organizers of the Women's March in Los Angeles did not grant Black Lives Matter participants speaking space at the rally for fear it would distract from their agenda, from the efficiency of their planned protest. Similarly, Tamika L. Carey's (2014, 2016) critiques frame Tyler Perry's works—generally popular with Black audiences—as anti-feminist: "Perry's representation of healing does not contribute to black women's liberation but rather points to a moment where black women's pain is a commodity" (2014, 1002). Black women are the identity group most often disadvantaged anytime efficiency rhetorics plague social justice efforts; it is for this reason that Natasha N. Jones, Laura Gonzales, and Angela M. Haas (2021), following the Combahee River Collective (1983), predicate an anti-racist future on the liberation of Black women. A complementary rhetorical approach is to privilege contextuality, following the community-focused research of Constance Haywood (2019, para. 7), who argues that "we should see reciprocity as always being contingent upon the desires and goals of the communities we engage."

Current-traditional understandings of efficiency—and we have seen many throughout this book—privilege expending the least effort possible. This understanding shapes narratives in the popular media every day. Efficiency is a framing concept even if it is not always evoked explicitly. (In fact, part of its power as a framing concept is precisely that it is invoked explicitly so rarely—even less often than objectivity or neutrality.) It shapes narratives about the DHD in a number of ways. For example, on a macro level, efficiency works to cover over bodies while making ecology and economic concerns more apparent. Why is it okay to talk about fish and birds but not people? What is the value system we are buying into that allows this to happen? The pelicans in the Gulf after DHD "became traumatic, charismatic stand-ins for a microbial and cellular catastrophe whose temporal and physical dimensions we are ill equipped to imagine and the science of which we do not adequately understand" (Nixon 2011b, 269). As a metaphor, those same pelicans might stand in for human bodies—or they might render them less apparent, creating an ecology-focused frame.

Apparent feminist efficiency frames would shift the concepts that are possible to talk about. And, to be clear, a vast number of frames exist on a continuum that might speak more or less to apparent feminist values. Some media outlets, especially smaller and niche outlets, have made attempts to shift efficiency frames. In the same way, Rob Nixon (2011b, 270) demonstrates a binary, arguing that we "seesaw between two narrowly defined definitions of risk: the risk of relying on foreign oil and the risk of domestic drilling." He argues instead for a third option that has been "interminably deferred": post-hydrocarbon possibilities. What I find of import here, beyond Nixon's argument for sustainable energy, is the way he uses slow violence as a lever with which to shift the available possibilities. By calling attention to articulations of our current risk scenario as hopelessly mired in dichotomy, he helps to imagine other possibilities. At the same time, this imagining is most effective if it happens in language that is legible to stakeholders. As Josephine Walwema and Felicita Arzu Carmichael (2021) show in their study of US immigration legislation and technical communication job descriptions, and as Michelle P. Covi, Jennifer F. Brewer, and Donna J. Kain (2021)[4] show after conducting interviews regarding perceptions of sea level rise, we can affect change efficiently by using language that actors in the current system recognize.[5] We can use apparent feminist efficiency framings in the same way—not as a method for inventing or investing in the other side of a polarity but as a method for thinking about boundless other radical but legible possibilities.

Technical communication scholars are no strangers to reframing; we can even already find examples of implicit reframing in reference to the DHD in our field literature. Daniel P. Richards (2017, 150) examines the ways post-disaster narratives related to the Deepwater Horizon Disaster use a common topos of cause and effect as their primary frame, masking "the complexity of the human-technological relationships." Richards aims to "think about the ways in which accident reports as a public mode of technical communication have the opportunity to help environmental rhetoric move beyond stifling and unproductive invention strategies pertaining to causality in disaster and toward articulations of causality that provide a more productive pathway for future prevention" (151). Deborah M. Welsh (2014) explores the ways various types of opening statements (legal and non-legal) co-construct and "control the version of the reality of" the Deepwater Horizon Disaster.

Controlling reality is, of course, what this entire book is about. What crises are real, and what are the criteria by which we say so? What criteria are efficient for our purposes? As we learned at the very beginning of chapter 1, it is necessary to have data to compare new data against: "Contemporary revolutions are indexed on the immediately prior state of the system" (Baudrillard 1976, 24). Without an original data set, we are left to construct one from our memories and feelings. (Locals in Mobile say there is a nitrogen dead zone in the Gulf that the agriculture lobby doesn't want us to know about. Define dead zone. How dead? Compared to when?) This leaves us vulnerable to having our perspectives shifted.

Critical and feminist scholars who have theorized shifts in perspective based on ephemeral "facts" often talk about temporality, as I showed when theorizing slow crisis. James M. Broadway and Brittinay Sandoval (2016, para. 4) say, "Our retrospective judgment of time is based on how many new memories we create over a certain period." Lauren Berlant (2011, 99) had previously captured this distillation of more recent science: "I recast this situation within a zone of temporality marked by ongoingness, getting by, and living on, where structural inequalities are dispersed and the pacing of experience is uneven and often mediated by way of phenomena that are not prone to capture by a consciousness organized by archives of memorable impact." She further extrapolates on her concept of slow death by pointing out that crisis always presupposes change, and thus crisis rhetoric is always "neither a state of exception nor the opposite, mere banality, but a domain where an upsetting scene of living is revealed to be interwoven with ordinary life after all" (102). Crisis as ordinary is a framing that might allow us to talk about a number of new approaches, including approaches that emphasize being proactive in the midst of ongoing crises.

To be clear, none of what I have argued here is objective or neutral. I don't claim that it is, and if I did you should be suspect. Nothing is neutral; nothing is objective. Efficiency frames are "inevitably part of an argument about classification, causality, responsibility, degeneracy, and the imaginable and pragmatic logics of cure" (Berlant 2011, 103). While Berlant is talking about epidemic rhetorics here—thus her use of the term *cure*—I take a wider view of it, considering that cure might mean the end to or amelioration of (perceived) crisis.

And yet, the end of crisis is also a subject of contention and a particularity that apparent feminisms' *slow crisis* might ask us to trouble. In *Unruly Rhetorics* (Alexander, Jarratt, and Welch 2018), Dana L. Cloud (2018, 41) argues that the Texas #feministarmy of 2013, raised in response to Texas legislators' abhorrent behavior toward women,[6] lost its power when "leaders of mainstream electoral politics intentionally un-occupied the Capital, taking the demonstrators away from the heart of their power into a domain controlled by others." Protestors were redirected to attend a Democratic Party rally at another site with the hopes of translating the energy of the rally into increasing voter turnout. These efforts failed, in large part because they did not recognize the (unruly) nature of the gathering. These weren't people who had shown up to play by the rules. They were tired of playing by the rules when the opposition refused to do so. We might understand this loss of power as constituting either the end of a perceived crisis, in that the protestors dissipated when the vote was invalidated, or the beginning of one, in that the protestors did not maintain the energy Democrats had hoped to channel. The subject of the affect in question determines the boundaries of the crisis. The manner in which this situation could be construed as crisis depends on one's specific perspective, time, and focus.

One potential apparent feminist reframing of the DHD might see efficiency in terms of happiness. Sara Ahmed (2011, 163) says happiness "involves a specific kind of intentionality, which I would call 'end oriented' . . . Classically, happiness has been considered as an ends rather than as a means." What might it mean to take happiness as a means? In the context of the Deepwater Horizon Disaster's aftermath in Mobile, it might mean honoring residents' choice to focus on economic recovery. Ahmed also spends much time discussing the "promissory logics of happiness," which suggests an underlying economic frame to this analysis—or perhaps this is a reframing of economy writ large. Ahmed is concerned with how feelings are directed toward objects situated in time; objects in the present, objects in the future, and so on: "If we pass happy objects around, it does not necessarily mean we are passing a feeling, but rather the promise of a feeling, or even the feeling of a promise" (164) Queer time, then—a necessity for slow crisis, remember—is about happiness or the potential for happiness (the two really being the same thing): "We would not wait for things to happen. To wait is to eliminate the hap by accepting the inheritance of its elimination. You make happen. Or you create the ground on which things can happen in alternative ways" (178).

This making that Ahmed discusses is an important moment of forward movement for apparent feminists who are seeking to reframe present and future conversations, but it also opens the door to reframings of the past. Elsewhere, I have discussed the efficiencies involved in seeing past/published/competed work that does not identify as feminist as apparently feminist. I have wrestled with the politics and ethics of labeling a work or a person as feminist if that person/author has not identified as such. My personal brand of feminism does not tend toward thrusting labels on people, and I see self-identification as a basic human right. However, rhetorical scholarship can provide us with an opening to interpret, or reframe, past work in apparently feminist ways. If the author is dead, as Roland Barthes (1977) suggests, we should (and perhaps cannot help but) use the work in question as we please. We might put limits on this approach for ourselves; perhaps we label work apparently feminist only under certain conditions (if it does not explicitly disavow feminisms or if it doesn't take up sexual politics explicitly, for example). We might also recognize that calling something apparently feminist could mean it is a complementary approach for our own feminisms, even if that does not make the work itself *actually* feminist. One excellent example of this would be womanism. I would not seek to label a womanist scholar as feminist, but apparent feminism might help me recognize the ways this scholar's work is complementary to my own and vice versa.

The question of feminist identity and time is a sticky one, as is the question of the progression of time alongside almost any social issue. Elizabeth Grosz (2004, 254–255, original emphasis) provides an avenue for resituating this quandary, and she does it by shifting an efficiency frame:

> Instead of the past being regarded as fixed, inert, given, unalterable, even if not knowable in its entirety, it must be regarded as inherently open to future rewritings, as never "full" enough, or present enough, to propel itself intact into the future . . . The past, in other words, is always already contained in the present, not as its cause or its pattern but as its latency, its virtuality, its potential for being otherwise. This is why the question of history remains a volatile one, not simply tied to getting the facts of the past sorted out and agreed on. It is about the production of *conceivable futures* . . . This indeed is what I understand feminist politics—at least at its best—to be about: the production of futures for women that are uncontained by any of the models provided in the present.

By shifting the god term from past to future—by insisting that history is about the future rather than the past and that we can do more efficient historical work by looking to the future—Grosz makes room for alternative understandings of feminist politics. She points out that, then, "history is not the recovery of the truth of bodies or lives in the past; it is the engendering of new kinds of bodies and new kinds of lives" (255). What kinds of lives might we imagine if we see the most efficient histories as those that most directly shape the future?

For the purposes of imagining a new efficiency model for feminist politics, Grosz's work can help extend the concept of slow crisis and its reach.[7] Grosz (2004, 249) argues that "time lends itself to spatialization, enumeration, and geometrical modeling, although these processes lose what is essential to temporality: its dynamic movement." If we focus instead on dynamic movement as an efficient value, we can see radical new possibilities for response to and preparedness for crisis events, including the DHD. Imagine if dynamic movement—positivity, constant reinvestment—were the efficiency principle underlying the DHD response. When I visited the Mobile County Health Department, a dynamic movement efficiency framing would have meant that administrators there saw me as a resource rather than a threat; they might have made use of my background in reporting to help reach their constituents, or my research skills to help them compile information, or my investment in their community to make better connections to academic institutions and the privilege they wield.[8] If dynamic movement were an efficiency principle espoused by the federal government, they might have tried more and different ways to counteract the effects of the spill, drawing more heavily on Indigenous and local knowledges than on chemical dispersant. If dynamic movement were an efficiency principle for the corporate actors involved, the value of profits might be imagined very differently, freeing up additional funding for dealing with a disaster of such proportion. If we take dynamic movement more literally as an efficiency frame, we might have found environmental scientists with the resources to study and interest in studying literal movement—migratory patterns of all kinds of animals (including humans)—instead of focusing on fatalities alone in the wake of the spill.

One thing that is clear is that science itself is built in particular ways, with particular efficiencies in place that do not allow for efficiency framings like dynamic movement. Grosz (2004, 246) theorizes that science's tendency to reduce "time to modes of the spatial and the countable has left science with a number of paradoxes, potentially insoluble problems, problems that science cannot in the long run ignore and that threaten to transform the nature of the sciences themselves as they continue to be drawn again and again to these inherent and irresolvable problems." I find stagnancy to be the more likely outcome and the greater threat to scientific efficacy. The problems of inflexibility that Grosz and other feminist theorists have exposed in scientific reasoning (see Crasnow, Wylie, Bauchspies, and Potter 2018; Derkatch 2016; Harding 1986) are more likely to hold scientific inquiry rigid, and the threat of rigidity is why feminist perspectives are needed. Without the ability to react to changing efficiency frames, the nature of the sciences themselves will cease to be useful in shaping our worldviews and our responses to natural phenomena—whether we realize it or not. We need feminist methodologies to take on the task at hand: "The task is not so much to plan for the future, organize our resources toward it, to envision it before it comes about, for this reduces the future to the present. It is to make the future, to invent it" (Grosz 2004, 261). We need apparent feminism to construct a future that is equitable and responsive, a future that is built on an efficiency model in which we focus on result gain as much as effort expended and make apparent who is receiving results and who is investing their efforts.

As we have seen, efficiency can be employed to overwrite health concerns. This happens because slow crisis is not recognizable to us *as* crisis based on our framing of efficiency. The only exceptions to attention to health concerns come when fast crises present themselves. These occur in instances where we are able to recognize vulnerable populations, such as embryos and fetuses, for which development is impacted. Thus, an apparent feminist efficiency reframing might be one that helps us to recast time—as I have done here with this concept of slow crisis as it extends toward the future—in ways that make crises, perhaps especially crises of human health, more apparent.

SITES OF ANALYSIS: A FINAL NOTE ON HEALTH

The (lack of) apparency of human health crises is an issue with reach far beyond the DHD. As I write this conclusion, new spikes in Covid-19 infection rates and political fracas about whether to reopen states as a result are topics very much in the news. While human health has been an explicit focus in the instance of this global pandemic and thus this case does not require an apparent feminist analysis as urgently as other circumstances, it *is* an excellent example of the overt ways some communicators privilege economy over health. Many conservatives have been forthright about saying that a certain percentage death toll or infection rate is simply the price we must pay for reopening the economy. In this instance, the connection between economics and human health is explicit. In other cases, however, the connection exists but is carefully covered over by mutual agreement.

Thus, it is important that apparent feminism and slow crisis can reveal health crises in more mundane contexts—contexts where we would not imagine crises occurring but where they do everyday. For example, many studies have shown that women use more healthcare services than men and tend to be the intermediaries between the medical field and their families—suggesting that human health is a feminist issue in the sense that men are socialized not to manage their health, women are socialized to take on the burden of *everyone's* health, and non-binary people often find healthcare inaccessible altogether. Klea D. Bertakis and her coauthors (2000) found that women were more likely to visit their primary care clinic and to seek diagnostic services than were men, even after controlling for health status, socio-demographics, and clinic assignment. Each of these socialized gender roles could constitute a crisis. Both Paula Johnson (2013) and Alana Baker (2017) have shown that "gold standard" treatments are based on "standard" bodies, which has typically meant cishet white male bodies without comorbidities; this results in an enormous lack of data about how to treat "nonstandard" bodies. This is certainly a crisis. Indeed, standardized treatment—or even treatment from experience (as opposed, perhaps, to a position of apparent feminist disidentification)—can result in missed or incorrect diagnoses. Kerri K. Morris (2020) raises concerns about exclusionary practices as a result of sociocultural understandings of illness in "Women and Bladder Cancer: Listening Rhetorically to Healthcare Disparities," wherein she relates her own experience of being—finally—diagnosed and then providing her physician with a mapping of female anatomy, since their exam room had previously lacked charts for educating women about bladder cancer.

Other crises may seem less urgent but have deep impacts on our worldview. Several years ago, I launched a large digital survey on perceptions of medical imaging. Between November 15, 2016, and April 12, 2017, I collected 162 responses. Of the 153 participants who responded to a question about gender, 0 participants identified as non-binary, 37 participants identified as male, and 116 (75.82%) participants identified as female. This alone suggested to me that women are more conditioned to be generous with their healthcare information. Ultimately, the survey and subsequent contextual analysis showed differences in how often women are the subjects of medical imaging; I further argue that this has implications for—as the survey response showed—our understandings of privacy as well as our political agency, to say nothing of our understandings of efficiency frames when faced with health problems.

Finally, though I lack the space to engage in a larger survey of technical communication's treatment of human health, recent publications on the state of this subject bear mentioning. A recent special issue of *Technical Communication Quarterly* (Frost et al. 2021) on "Unruly Bodies, Intersectionality, and Marginalization in Health and Medical Discourse" addresses technical communicators' approaches to rhetorics of health and medicine. When conducting research before proposing this special issue, a co-editor and I found that women's healthcare research is most often about reproduction. Indeed, treatments of women's reproductive health have increased in recent years relative to non-reproductive and holistic approaches to women's health. Using medicalrhetoric.com as our main bibliographic resource, we found a dearth of research focused on gender/queer and women's health issues (23.2% and 15.2 % of total intellectual work in the field of rhetorics of health and medicine, respectively); also telling is that of that those scholarly works about women's health, 75 percent are about women's health in a reproductive capacity. In other words, we saw a crisis in the ways technical communicators think about women's health, and we hope the special issue might interrupt notions of women's health as always already related to reproduction—and the efficiencies that underlie this notion.

DOING THE WORK: A BEGINNING

Apparent feminists must actively engage in work that makes more apparent existing and potential sites of resistance. We might label the practices of revising origin stories and making bodies apparent to provide new entry points for intervention—as I have done in this chapter—in many ways. The approach I have employed in this chapter is subversive historiography, which hooks (2004, 399–400) says "connects oppositional practices from the past with forms of resistance in the present, thus creating spaces of possibility where the future can be imagined differently—imagined in such a way that we can witness ourselves dreaming, moving forward and beyond the limits and confines of fixed locations." Apparent feminists must search out examples of past resistance, draw on those examples to revise histories, and use the new entry points they uncover to build toward future transformation. In other words, I suggest that hooks's subversive historiography is one way of thinking about potential apparent feminist approaches to revising the reality of technical communication as a discipline.

Most important of all is the necessity for apparent feminists to be willing to run the risks associated with feminisms. Despite the dangers inherent in doing so, we must be assertive and tactical in maintaining feminist apparency in order to intervene in oppressive hegemonic formations that hide behind rhetorics of post-feminism. To draw on hooks (1989, 9) again for an illustrative example, she discusses how she chose the name bell hooks—drawn from her mother's and grandmother's names and intended to intentionally de-center herself—"to construct a writer-identity that would challenge and subdue all impulses leading me away from speech into silence." For hooks, talking back "meant daring to disagree and sometimes it just meant having an opinion"; it is a "courageous act—an act of risk and daring" (5). Having an opinion—making embodied perspectives apparent—means being willing to intervene, to introduce challenges, and sometimes to be a problem for privileged rhetors. Moving ourselves to speech and action is also a method of making certain injustices apparent despite the powerful institutions that support and maintain those injustices while attempting to delegitimize the bodies we make such critiques from.

Finally, it is desperately important for apparent feminists to recognize, draw upon, and contribute to the connections that sustain us in our pursuit of social justice. These connections are interdisciplinary, transnational, and transcultural. As just one example, feminist and gender studies traditions have used many organizing metaphors that cross all kinds of supposed disciplinary, national, and cultural boundaries. Some of those metaphors include "Gloria Anzaldúa's borderlands, Jacqueline Royster's stream, Cheryl Glenn's silence, Krista Ratcliffe's listening, Nedra Reynolds's geographies, Wendy Hesford and Wendy Kozol's advocacy, Eileen Schell and K. J. Rawson's motion, and Lindal Buchanan and Kathleen J. Ryan's walking and talking" (Flynn, Sotirin, and Brady 2012, 2), as well as Elizabeth A. Flynn, Patricia Sotirin, and Ann Brady's resilience, Christina Sharpe's (2016) wake, Chandra Talpade Mohanty's (1988) gaze, Winona LaDuke's (1999) relations, and Francesca Bray's (1997) and Jack Halberstam's (2005) place and time.

These interdisciplinary connections are important for the sake of determining what we mean when we employ apparent feminist *efficiency*; who we are working for, in what context, and within what time frame will be constant problems for apparent feminist technical communicators. To move toward social justice, we must first welcome and make space for the presence and participation of allies. The sustainability of better conditions for one group of people depends on the sustainability of similar conditions in other parts of the system; "wealthier nations will not be secured financially or geopolitically if the poor are not part of a modern, global, and capitalist economy" (Dingo and Scott 2012, 2).[9] That is, we must imagine the ultimate apparent feminist efficiency as a scenario in which socially just economies are created and sustained across disciplinary, national, and cultural borders.

Indeed, community—and paradigms that privilege being in community—may be the most important element of such an apparent feminist approach. Being in community with others is often where I find paths forward. For example, I recently helped facilitate a workshop for the annual meeting of the Association of Teachers of Technical Writing. My colleague Michelle Eble and I offered ideas for how to make shifts in graduate programs' structures so they are better designed for supporting BIPOC students. Because of our own context and positionality, we offered a lot of ideas for action items white women administrators might perform as they seek to support Black students and especially Black women, and we asked participants to help by contributing knowledge of their own situatedness and context. One set of questions we specifically asked attendees to grapple with was: where are the roadblocks in your program? Where do students, especially BIPOC students, struggle or get stuck? Where/when do people leave the program? Why, and how do you know? As is common in such presentations and workshops, we were quickly awash in stories of such roadblocks, all saturated in a culture of systemic injustice that complicated any efforts at improvement. We found ourselves frustrated with and stymied by the institution that is the university, and we recognized that the university was not built for anyone in that session. Our session, "Building and Sustaining Inclusive Graduate Programs," appeared to be attended entirely by women-identified people and BIPOC, and all of us understand that "many of our elite university buildings were physically built by enslaved people on stolen land so that white men would be able to train the next generation of those who belong: white men" (García Peña 2022, 15) and that the absence of white men in the session was not coincidental.

But sometimes a trajectory can be born from collective frustration. At one point, Angela Haas asked what it would look like to build a "coalitional abolitionist anti-colonial university," and Natasha Jones asked, "what does abolitionist education look like for us?" Mari Ramler recommended the book *Community as Rebellion*, wherein Lorgia García Peña (2022) discusses the founding of Freedom University—which was created explicitly for undocumented students and thus subverts many traditional university structures.

The session continued in a number of directions, one of which involved theorizing metaphors for ways we can build social justice spaces within the existing university system. We talked about building an oasis, or a bubble—a place where students can exist in community with one another, bound by a shared commitment to social justice and to learning. I found myself thinking about the notion of a "bubble" in a new way. Throughout this text, and following a long tradition of work in the field, I use the language of "the turn." This metaphor is used to describe a field taking up a new orienting concept in a substantial way: the cultural turn, the social turn, the social justice turn. Depending on our orientations, certain turns may matter more to us than others. They may impact our most *efficient* approaches to engagement in the field differently, depending on our embodiments and commitments. Indeed, certain turns may be more and less readable or apparent to us; a turn that defines my work may not mean much to my colleague. I tend to find anything my colleagues would term a "turn" to be interesting and valuable, but I also remember the feeling of overwhelm as a graduate student and early-career scholar, when I felt as though I was "turning" every which way and was terrified I would miss something vital that was hovering on my periphery. In moments of frustration, I imagined myself spinning senselessly, paralyzed by my inability to attend to everything that's important. The metaphor of the bubble, though, can perhaps help overcome that type of paralysis. If we imagine a bubble nest—a mass of connected bubbles with no center—we have a metaphor for organizing the field's turns that is malleable, permeable, temporary, and structured but center-less. When you figure out that you've missed something, you can add a bubble. That bubble won't change the whole structure immediately, but it could in the long term: "Anti-racist work, and more specifically pro-Black and liberatory work, should be preceded by the necessary research to do said work" (Jones, Gonzales, and Haas 2021, 31). As this work continues iteratively, the whole structure of a person's knowledge can change. Perhaps most of all, this metaphor mirrors a community since it is based on interconnection.

Writing studies and rhetorics of health and medicine are just now echoing what ethnic studies has been telling us: that rebellion is built on community. In the aforementioned book *Community as Rebellion*, García Peña (2022, xiv) explains that it was her caregivers—her grandma and aunties—who "made my rebellion possible." Later, her experience on fellowship at Syracuse University meant that "for six months, I was surrounded by the care and mentorship of feminist and BIPOC scholars, including Chandra Talpade Mohanty, Linda Carty, and Silvio

Torres-Saillant. They held my hand and helped me through my insecurities . . . gave me sincere and thoughtful feedback . . . Their accompaniment showed me that, indeed, another way is possible" (48). García Peña has worked to replicate this experience for others in helping to create Freedom University.

García Peña (2022, 19–20) describes an approach wherein we do not (futilely) demand that the institution love us back but rather that we "take its resources and structures and repurpose them to create freedom spaces, freedom schools, and liberation moments within and through its violent exclusion." Freedom University, located in Atlanta, "provides tuition-free college preparation classes, college and scholarship application assistance, mental health and legal support, and social movement leadership development for undocumented students" (Freedom University: About, para. 1). The university was created as a result of the state of Georgia's banning of undocumented students from some of its top public universities. Freedom University is born of the labor of many activists who have operated from García Peña's (2022, 31) philosophy that "we must rebel by creating communities of freedom within and outside the institution, by reaching out to others and forming concrete plans to sustain our work and our lives."

Apparent feminism—like rhetorical resilience—does not assume that its subjects "can marshal resources and have access to forums for public action" but instead values the possibilities for productive work coming from "a place of struggle and desire" (Flynn, Sotirin, and Brady 2012, 7). Freedom University is an ambitious (and risky) project, beyond the reach of many of us, but its scope doesn't diminish the importance of smaller apparent feminist ventures. Valuing sites of oppression as places for productive work is precisely the goal facilitated by using apparent feminism to rearticulate the origin stories of technical communication in this chapter. Apparent feminism recognizes the impossibility of convincing every audience of the value of feminisms; apparent feminism also recognizes the importance of social justice work produced by those who may not have the agency—whether their constraints are theological, cultural, national, political, legal, disciplinary, physical, intellectual, or emotional—to mobilize and maintain a feminist identity.

Those who do have agency, then, have an increased obligation. They have a responsibility to the rhetorical velocity of the work I discuss above; those with privilege are responsible not only for doing their own apparent feminist work, but they also must be responsible for helping make apparent the work of those with less power and agency than themselves. This obligation speaks to social justice, but it also speaks to efficiency. In many cases, for example, it may be more efficient for someone embodied as a white male to do apparent feminist work in the academic discipline of technical communication (and certainly in many other contexts as well). White men doing this kind of work may not encounter the same risk as those embodied in other ways; white men doing this kind of work may also be heard more easily by the larger community. Further, as a cis white woman, my privilege includes some measure of access to greater efficiency in this regard than is the case for some of my colleagues embodied in other ways. In doing my own apparent feminist scholarship—throughout this book and elsewhere—I am cognizant of drawing on and speaking back to work by those who may not have the same agency I do. I hope to make that work more apparent to technical communication scholars. I—and many others—have also benefited from the generosity of scholars who have used their greater agency to make our work more apparent.[10] Whatever our relative privilege, my hope is that investing in a sense of apparency toward new understandings of efficiency can promote social and environmental justice. Our work, our relationships, our bodies, our health, our very lives depend on it.

NOTES

PREFACE: ON POSITIONALITY AND INCLUSION

1. In some ways, then, an apparent feminist approach's willingness to let go of terms may seem fairly radical. Sometimes radical is necessary; I return to this later.
2. This distinction ("this country") is about the limits of my own expertise, not about the limits of racism.
3. Pouncil and Sanders (2022, 294) also offer a useful set of questions for non-Black technical communicators to engage with as part of a critically reflective praxis:

    What does it mean to meaningfully and accountably labor toward coalition that implicates your identities, and those you share identities with, toward interacting with Black people seeking coalition with you?
    Consider what it may mean to Black folks to interact with you while you are learning how to be in coalitional alliance with them.
    How do you develop critical reorientations to the work of TPC [technical and professional communication], particularly those that challenge and unsettle claims to objectivism, neutrality, and a unified truth while not burdening or harming your Black colleagues?

4. That is, objectivity as it is often imagined: a neutral perspective untainted by cultural or social associations.
5. I have followed the authors' suggested citation format as indicated in the document itself, but my stance on positionality and my commitment to fair citation practice dictate that I attach their names to their work. Contributors to this project included Lauren E. Cagle, Michelle F. Eble, Laura Gonzales, Meredith A. Johnson, Nathan R. Johnson, Natasha N. Jones, Liz Lane, Temptaous Mckoy, Kristen R. Moore, Ricky Reynoso, Emma J. Rose, GPat Patterson, Fernando Sánchez, Ann Shivers-McNair, Michele Simmons, Erica M. Stone, Jason Tham, Rebecca Walton, and Miriam F. Williams.

CHAPTER 1: FEMINIST TECHNICAL COMMUNICATION

1. See Connors (1982) and McDowell (2003) for traditional histories of the discipline of technical communication.
2. Troubling the origin stories of rhetoric as a discipline (which often posit its beginning in ancient Greece, although other possibilities—like Mesopotamian traditions—do exist) is an equally important project that has already been taken up by a number of scholars (Crowley and Hawhee 2012; Glenn 1994; Jarratt 1998, 2002).
3. Haas (2007) specifically discusses the rhetorical tradition of wampum belts and strings as digital cultural-rhetorical artifacts.
4. Prostitution is a common way for single women—especially single women with children—in Masesse, Uganda, to earn income. Juliette worked as a prostitute to feed her four children after her husband left the family.
5. Convened by 2019 CCCC chair Vershawn Ashanti Young, whose call for papers for the conference unapologetically eschewed traditional academic register and

insisted on centering Black technical expertise without marking this practice as anything unusual.
6. Disciplinary friction is important both in the sense that it is a vital part of our history and also in the sense that without friction, we struggle to move forward. Friction provides something to push against, a toehold on which to lever ourselves toward something better (Mckoy, Shelton, Davis, and Frost 2022).
7. Connors's history of technical communication spends considerable time on the rise of engineering schools—the traditional arc of histories told about technical communication—although he first acknowledges that technical writing has existed as long as tools have existed.
8. As one example, when I advocated that we use a hire the college had afforded us to recruit someone with an English studies degree who would be situated to teach both literature and rhetoric, a colleague (who I very much like and respect) who specializes in literature responded, "If we can do that, then why would we hire a literature specialist ever again?" My hope is that this sort of friction can move us to new ideas about hiring across disciplines; this remains to be seen.
9. To be sure, it would also be a grave error to treat first-wave feminism as a monolith—although that conversation is beyond the scope of this book.
10. Mary Lay Schuster (2015) tells the story of how this article came about, along with other related history of women in the field of technical communication.
11. Charlotte Thralls, also the co-editor (with Mark Zachry) of *Communicative Practices in Workplaces and the Professions: Cultural Perspectives on the Regulation of Discourse and Organizations* (2007), was editor of *JBTC* when Lay's article was published. Thomas Kent was the editor when the 1992 special issue came out. Isabelle Thompson (1999, 163) notes in her survey of technical communication articles on gender and feminism that "the journals publishing the most articles about women and feminism are currently edited by women," and she shows that *JBTC* and *TCQ* easily outpaced the other journals in her corpus (*IEEE Transactions on Professional Communication, Journal of Technical Writing and Communication, Technical Communication*) in terms of the percentage of articles published about women and feminism.
12. The cultural turn discussed in this issue foreshadows my point about feminisms being subsumed under cultural studies; however, at the time this issue was published, feminism was still explicitly named as an integral part of a cultural studies approach.
13. To be clear, while I favor apparency when possible, apparent feminism asks its practitioners to make informed choices in considering and encouraging apparency. In some cases, being apparently/explicitly feminist may not be the best rhetorical move. A savvy apparent feminist communicator must be able to assess the rhetorical landscape and determine whether it is favorable toward, persuadable of, or hostile to (various, including feminist) rhetorics of difference. Stephanie Kerschbaum (2014) offers a thorough analysis of contextual ways of (de)valuing difference and how to parse institutional conversations around difference and diversity.
14. For further reading on feminist intellectual and activist transmigrations, see Haas (2008a). She suggests that we improve on Mary Louise Pratt's (1999) concept of contact zones by invoking "transmigrations" instead. This move is born out of the fact that "the legacy of the contact zones on the colonial frontier substantiates that such zones were often not safe, respectful, or reciprocal for indigenous peoples" (Haas 2008a, 56). Transmigratory practices, in contrast, are "dedicated to respectful and reciprocal dialogue between and across culturally-specific dig/viz rhetorics" (57).
15. See Scott (2004), Sauer (2003), and Sandman (2020). for varying definitions of "risk," for example.

CHAPTER 2: APPARENT FEMINISMS

1. Visible rhetoric is a sub-discipline of rhetoric that focuses on the persuasive use of non-alphabetic modes of communication (e.g., the appropriation and captioning of ultrasound images to persuade pregnant people to carry pregnancies to term).
2. The overturning of *Roe v. Wade* in June 2022 precipitated a rush of abortion-related legislation, but little of it focused on ultrasound since states that previously required a persuasive vehicle can now ban abortion outright.
3. The term *people* is used purposefully here and throughout. While the most common sites of fetal ultrasound are the bodies of women, pregnant people may identify in a variety of ways. In addition, the partners of pregnant people are also victims of this kind of legislation. I use the term *people* to be inclusive of different groups affected and also to point out that feminisms work in service of more than just the group that self-identifies as women.
4. The term *social justice* refers to an obligation to widen our perspective when thinking about fairness in social situations. The term *social* indicates recognition of more than a single (kind of) subject, refocusing our energy on broader, more inclusive, and diverse applications of our existing understanding of the principle of justice.
5. Self-identification is a complex subject. Suffice it to say that identification, like apparent feminism, is always in flux (see Ratcliffe 2005).
6. For a detailed analysis of ethos in technical documentation, refer to Frost and Sharp-Hoskins (2015).
7. *Diversity*, like feminism, is a term laden with many possible connotations. Further, much like efficiency, it is a term that is sometimes invoked symbolically and without critical analysis (Ahmed 2012). I use diversity here to talk about the importance of considering cultural difference in technical audiences. While I focus here on gender-based and sex-based differences, this cultural diversity could (and should) also include ethnic, national, religious, philosophical, and political diversity—among other meanings of diversity that my own cultural positioning make unapparent at this juncture.
8. Some of these sources are dated, so they are articulating perspectives on the discipline as it looked at another point in time. I include these sources, as well as more current ones, to show that even as the field changes over time, some patterns, unfortunately, seem to be rather stable.
9. According to an interview in *Lambda Literary* (Sexsmith 2012), Halberstam said, "So some people call me Jack, my sister calls me Jude, people who I've known forever call me Judith—I try not to police any of it. A lot of people call me he, some people call me she, and I let it be a weird mix of things and I'm not trying to control it." I appreciate Halberstam's radically fluid approach to identity and have thus cited his work using the name that appears on that given published work.
10. Interestingly, N. Katherine Hayles's (2021) posthuman take on Covid-19 points to the oppositional evolutionary strategies of humans and viruses; namely, she juxtaposes humans' progressively more complex genetic codes with viruses' progressively simpler ones. Viruses follow something like the approach I allude to here, cutting away anything that isn't useful. Importantly, though, a virus-based value system is about replication above all else, which is the opposite of what I'm advocating. Hayles ultimately invites readers to avoid binaries and to think more deeply about relationships and interdependencies among humans, non-human species, and artificial agents.
11. Indeed, I have only mentioned three of the twenty-four states that have passed similar legislation (Frost 2013a).

CHAPTER 3: SLOW CRISIS

1. *Secured* may be an optimistic word, as the right to marry remains under threat.

CHAPTER 4: DISASTER

1. I did not ultimately participate in cleanup efforts because although some small tarballs were reported prior to my visit, significant quantities of oil did not come ashore until June 26—the day after I left Dauphin Island, Alabama.
2. My experience as a reporter comes from time spent in Adair County, Missouri, and Logan County, Illinois, each of which has a population of about 30,000. My experience in the rural South comes from having lived in North Carolina since 2013 and from my many experiences as a visitor/tourist in southern Alabama (Frost 2013b).
3. MCHD had eleven total locations at this time. I did not visit locations that catered to specialized populations, such those focused on HIV or clinics based in schools.
4. I have obscured this source's identity for their protection.
5. Suits must be filed within four years of the date of diagnosis. Note that diagnosis constitutes some recognition by a medical provider but does not confirm a causal relationship to the DHD.
6. Indeed, some research has shown that even the notion of "choice" is problematic, as it "advances White, monolingual feminism, while eliding and even erasing more precarious positionalities and perspectives" (de Onís 2015, 10).

CHAPTER 5: AN APPARENT FEMINIST ANALYSIS OF THE DEEPWATER HORIZON DISASTER

1. For more on how cultural and technological factors contribute to building ethos online, refer to Davis (2018). For a nuanced troubling of the term *professionalism* and all it encodes, refer to Hull, Shelton, and Mckoy (2020); Mckoy (2019).
2. BP sued Halliburton, a US contractor, for part of the estimated $42 billion for cleanup efforts. BP claims Halliburton did faulty cement work on the Deepwater Horizon rig (Rushe 2012). This lawsuit was situated in the historical context of US-based legislation requiring those who create hazardous waste sites to be responsible for subsequent harm. This act, the 1980 US Comprehensive Environmental Response, Compensation, and Liability Act, significantly impacts the practices of corporations dealing with hazardous materials in the US (Lundgren and McMakin 2009). Halliburton ultimately settled out of court for a reported $1.1 billion (Reuters Staff 2015).

CHAPTER 6: LOOKING FORWARD, LOOKING BACK

1. The code name Macondo was drawn from the name of the fictional city in the book *One Hundred Years of Solitude*—a portentous choice since that city was destroyed by a hurricane.
2. It was the University of Southwestern Louisiana when she earned her degree in 1966.
3. In recent years, as a result of national political shifts, digital humanities and technical communication scholars have begun to conduct more in-depth and targeted studies of the role of digital spaces and technologies in (mis)information campaigns (e.g., Dorpenyo 2019; Ridolfo and Hart-Davidson 2019).
4. Covi, Brewer, and Kain (2021)'s work is especially relevant given its foundation in environmental change. They found significant ideological differences between

communities and concluded that "in the social production of risk, receptivity to science-based policy is shaped by local understandings of agency and fatalism that are selective and contingent, molded by fluctuating social constraints and opportunities" (9). They advocate for communicators to consider local social-political realities as they attempt to convey biophysical information. In short, they conclude that ideology is a powerful driver when it comes to policy development, even more powerful than overwhelming scientific evidence.

5. While an analysis of the effects of the plain language movement is outside the scope of my project in this passage, it's worth noting that Jones and Williams (2017) have shown that a "human-centered approach" that utilizes plain language can promote equity in the context of interactions between corporate stakeholders and marginalized groups.

6. One thing Cloud does not mention and that many other accounts of the end of Wendy Davis's filibuster also fail to mention—through no fault of their own, as it is not well documented—is the intention of the Republican caucus that night to subvert democracy. After the spectators in the chamber prevented the vote from occurring before midnight, a cluster of US Senate Republicans gathered around the lectern and cast an illegal vote. Because this happened in full view of the YouTube livestream with hundreds of thousands of people, including me, watching—which the senators perhaps did not realize at the time—the vote was invalidated before the opening of business the next day.

7. Grosz (2004, 258) articulates feminism as "not an inevitable effect of patriarchy, but [something that] lies latent or potential within patriarchal relations."

8. My home institution, East Carolina University, was familiar to the ecologists I spoke with because of its resources and reputation related to marine ecosystems.

9. Dingo and Scott's argument as quoted here draws on Arjun Appadurai's (1996) development of the phrase "megarhetorics of global development."

10. Gerald Savage, in particular, is a role model (for me and others) who has used his agency as a white man in the technical communication community to efficiently increase the apparency of diverse scholars, users, methods, methodologies, and scholarship. His dedication to social justice is evidenced in his mentorship and his publications (Kynell-Hunt and Savage 2003; Savage and Mattson 2011; Savage and Matveeva 2011).

REFERENCES

Agboka, Godwin Y., and Natalia Matveeva, eds. 2018. *Citizenship and Advocacy in Technical Communication: Scholarly and Pedagogical Perspectives*. New York: Routledge.
Ahmed, Sara. 2006. *Queer Phenomenology: Orientations, Objects, Others*. Durham, NC: Duke University Press.
Ahmed, Sara. 2011. "Happy Futures, Perhaps." In *Queer Times, Queer Becomings*, edited by E. L. McCallum and Mikko Tuhkanen, 159–182. New York: State University of New York Press.
Ahmed, Sara. 2012. *On Being Included: Racism and Diversity in Institutional Life*. Durham, NC: Duke University Press.
Alexander, Jamal-Jared, and Avery Edenfield. 2021. "Health and Wellness as Resistance: Tactical Folk Medicine." *Technical Communication Quarterly* 30, no. 3: 241–256. https://doi.org/10.1080/10572252.2021.1930181.
Alexander, Jonathan, Susan C. Jarratt, and Nancy Welch, eds. 2018. *Unruly Rhetorics: Protest, Persuasion, and Publics*. Pittsburgh, PA: University of Pittsburgh Press.
Ali, Nujood, and Delphine Minoui. 2010. *I am Nujood, Age 10 and Divorced*. New York: Three Rivers.
Allen, Jo. 1994. "Women and Authority in Business/Technical Communication Scholarship: An Analysis of Writing Features, Methods, and Strategies." *Technical Communication Quarterly* 3, no. 3 (Summer): 271–292.
Allen, Laura L. 2022. "Handling Family Business: Technical Communication Literacies in Black Family Reunions." *Technical Communication Quarterly* 3, no. 3: 229–244.
Althusser, Louis. 2006. "Ideology and Ideological State Apparatuses." In *The Anthropology of the State*, edited by Aradhana Sharma and Akhil Gupta, 86–111. Malden, MA: Blackwell.
"Anti-Racist Scholarly Reviewing Practices: A Heuristic for Editors, Reviewers, and Authors." 2021. https://tinyurl.com/reviewheuristic.
Appadurai, Arjun. 1996. *Modernity at Large: Cultural Dimensions of Globalization*. Minneapolis: University of Minnesota Press.
Appadurai, Arjun. 2000. "Grassroots Globalization and the Research Imagination." *Public Culture* 12, no. 1 (Winter): 1–19. https://doi.org/10.1215/08992363-12-1-1.
Appleton, Nayantara Sheoran. 2018. "Feminist Commons and Techno-Scientific Futures." *Commoning Ethnography* 1, no. 1 (December): 142–151. https://doi.org/10.26686/ce.v1i1.4120.
Arizona Women's Health and Safety Act. 2012. Ariz. HB 2036 §36–449.01.
Baker, Alana. 2017. "Pragmatic Feminist Empiricism: An Original Analytical Framework for Technical Communication." PhD dissertation, East Carolina University, Greenville, NC.
Ballif, Michelle. 2000. *Seduction, Sophistry, and the Woman with the Rhetorical Figure*. Carbondale: Southern Illinois University Press.
Baniya, Sweta, Les Hutchinson, Ashanka Kumari, Kyle Larson, and Chris Lindgren. 2019. *Representing Diversity in Digital Research: Digital Feminist Ethics and Resisting Dominant Normatives*. Proceedings of the Annual Computers and Writing Conference, 2018. https://vtechworks.lib.vt.edu/bitstream/handle/10919/96779/Baniya-etal-2018-cw_proceedings-representing_diversity.pdf?sequence=2&isAllowed=y.

Banks, Adam J. 2006. *Race, Rhetoric, and Technology: Searching for Higher Ground*. New York: Routledge.

Banks, William P. 2003. "Writing through the Body: Disruptions and 'Personal' Writing." *College English* 66, no. 1: 21–40.

Barnett, Scot. 2012. "Psychogeographies of Writing: Marking Space at the Limits of Representation." *Kairos* 16, no. 3. https://kairos.technorhetoric.net/16.3/topoi/barnett/index.html.

Barthes, Roland. 1977. "The Death of the Author." In *Image / Music / Text*, 142–148. Translated by Stephen Heath. New York: Hill and Wang.

Baudrillard, Jean. 1976. *Symbolic Exchange and Death*. Translated 1993 by Iain Hamilton Grant. London: Sage.

Belenky, Mary Field, Blythe M. Clinchy, Nancy R. Goldberger, and Jill M. Tarule. 1986. *Women's Ways of Knowing: The Development of Self, Voice, and Mind*. New York: Basic Books.

Bellwoar, Hannah. 2012. "Everyday Matters: Reception and Use as Productive Design of Health-Related Texts." *Technical Communication Quarterly* 21, no. 4 (June): 325–345. https://doi.org/10.1080/10572252.2012.702533.

Berlant, Lauren. 2011. *Cruel Optimism*. Durham, NC: Duke University Press.

Bernard, Mary, ed. 1999. *Sappho: A New Translation*. Berkeley: University of California Press.

Bernhardt, Stephen A. 1992. "The Design of Sexism: The Case of an Army Maintenance Manual." *IEEE Transactions on Professional Communication* 35, no. 4 (December): 217–221. https://doi.org/10.1109/47.180282.

Bertakis, Klea D., Reza Azari, Lori J. Helms, Edward J. Callahan, and John A. Robbins. 2000. "Gender Differences in the Utilization of Healthcare Services." *Journal of Family Practice* 49, no. 2 (February): 147–152. PMID: 10718692.

Biesecker, Barbara A. 1989. "Rethinking the Rhetorical Situation from within the Thematic of Différance." *Philosophy and Rhetoric* 22, no. 2: 110–130.

Bitzer, Lloyd F. 1968. "The Rhetorical Situation." *Philosophy and Rhetoric* 1, no. 1 (January): 1–14. https://www.jstor.org/stable/40236733.

Blackmon, Samantha. 2007. "(Cyber) Conspiracy Theories? African-American Students in the Computerized Writing Environment." In *Labor, Writing Technologies, and the Shaping of Composition in the Academy*, edited by Pamela Takayoshi and Patricia Sullivan, 153–166. Cresskill, NJ: Hampton.

Blair, Kristine L., and Lee Nickoson, eds. 2018. *Composing Feminist Interventions: Activism, Engagement, Praxis*. Fort Collins, CO: WAC Clearinghouse.

Boellstorff, Tom. 2007. "When Marriage Falls: Queer Coincidences in Straight Time." *GLQ: A Journal of Lesbian and Gay Studies* 13, no. 2–3 (September): 227–248. https://doi.org/10.1215/10642684-2006-032.

Booher, Amanda K., and Julie Jung, eds. 2018. *Feminist Rhetorical Science Studies: Human Bodies, Posthumanist Worlds*. Carbondale: Southern Illinois University Press.

Bosley, Deborah S. 1992. "Gender and Visual Communication: Toward a Feminist Theory of Design." *IEEE Transactions on Professional Communication* 35, no. 4 (December): 222–229. https://doi.org/10.1109/47.180283.

Bosley, Deborah S. 1994. "Feminist Theory, Audience Analysis, and Verbal and Visual Representation in a Technical Communication Writing Task." *Technical Communication Quarterly* 3, no. 3 (March): 293–307. https://doi.org/10.1080/10572259409364573.

Bowdon, Melody. 2004. "Technical Communication and the Role of the Public Intellectual: A Community HIV-Prevention Case Study." *Technical Communication Quarterly* 13, no. 3 (November): 325–340. https://doi.org/10.1207/s15427625tcq1303_6.

Bowdon, Melody, and Russell Carpenter, eds. 2011. "Guest Editors' Introduction: Community Literacy and Digital Technologies." *Community Literacy Journal* 6, no. 1: 1–4.

Boyer, Evelyn P., and Theora G. Webb. 1992. "Ethics and Diversity: A Correlation Enhanced through Corporate Communication." *IEEE Transactions on Professional Communication* 35, no. 1 (March): 38–43. https://doi.org/10.1109/47.126938.

Brady Aschauer, Ann. 1999. "Tinkering with Technological Skill: An Examination of the Gendered Uses of Technologies." *Computers and Composition* 16, no. 1 (February): 7–23. https://doi.org/10.1016/S8755-4615(99)80003-6.

Brasseur, Lee E. 1993. "Contesting the Objectivist Paradigm: Gender Issues in the Technical and Professional Communication Curriculum." *IEEE Transactions on Professional Communication* 36, no. 3 (September): 114–123.

Brasseur, Lee E. 2005. "Florence Nightingale's Visual Rhetoric in the Rose Diagrams." *Technical Communication Quarterly* 14, no. 2 (April): 161–182. https://doi.org/10.1207/s15427625tcq1402_3.

Bray, Francesca. 1997. *Technology and Gender: Fabrics of Power in Late Imperial China.* Berkeley: University of California Press.

Broadway, James M., and Brittiney Sandoval. 2016. "Why Does Time Seem to Speed Up with Age?" *Scientific American.* https://www.scientificamerican.com/article/why-does-time-seem-to-speed-up-with-age/ https:doi:10.1038/scientificamericanmind0716-73.

Brown, Claire Damken. 1993. "Male/Female Mentoring: Turning Potential Risks into Rewards." *IEEE Transactions on Professional Communication* 36, no. 4 (December): 197–200. https://doi.org/10.1109/47.259958.

Brunner, Diane D. 1991. "Who Owns This Work? The Question of Authorship in Professional/Academic Writing." *Journal of Business and Technical Communication* 5, no. 4 (October): 393–411. https://doi.org/10.1177/1050651991005004004.

Bureau of Safety and Environmental Enforcement. 2019. "BSSE Well Control Rule 2019." https://www.bsee.gov/guidance-and-regulations/regulations/regulatory-reform/bsee-well-control-rule-2019.

Burke, Kenneth. 1945. *A Grammar of Motives.* Englewood Cliffs, NJ: Prentice-Hall.

Butler, Judith. 1990. *Gender Trouble: Feminism and the Subversion of Identity.* New York: Routledge.

Butts, Shannon, and Madison Jones. 2021. "Deep Mapping for Environmental Communication Design." *Communication Design Quarterly* 9, no. 1 (March): 4–19. https://doi.org/10.1145/3437000.3437001.

Byrd, Antonio. 2019. "Between Learning and Opportunity: A Study of African American Coders' Networks of Support." *Literacy in Composition Studies* 7, no. 2: 31–55.

Byrd, Antonio. 2022. "Black Professional Communicators Testifying to Black Technical Joy." *Technical Communication Quarterly* 3, no. 3: 298–310.

Cagle, Lauren, Michelle Eble, Meredith Johnson, Nathan R. Johnson, Liz Lane, Temptaous Mckoy, Kristen R. Moore, Enrique Reynoso Jr., Emma Rose, Gpat Patterson, Fernando Sánchez, Michele Simmons, Erica Stone, Jason Chew Kit Tham, and Rebecca Walton. 2021. "Participatory Coalition Building: Creating an Anti-Racist Scholarly Reviewing Practices Heuristic." In *Proceedings of the 39th ACM International Conference on Design of Communication (SIGDOC '21),* edited by Elizabeth Lane and Halcyon M. Lawrence, 281–288. New York: Association for Computing Machinery. https://doi.org/10.1145/3472714.3473654.

Cammack, Becca. 2015. "Communication Practices by Way of Permits and Policy: Do Environmental Regulations Promote Sustainability in the Real World?" In *Communication Practices in Engineering, Manufacturing, and Research for Food and Water Safety,* edited by David Wright, 129–144. IEEE PCS Professional Engineering Communication Series. Hoboken, NJ: Wiley.

Campo-Engelstein, Lisa. 2020. "Reproductive Technologies Are Not the Cure for Social Problems." *Journal of Medical Ethics* 46, no. 2 (January): 85–86.

Cannon, Peter, and Katie Walkup. 2021. "Re/producing Knowledge in Health and Medicine: Designing Research Methods for Mental Health." *Technical Communication Quarterly* 30, no. 3: 257–270. https://doi.org/10.1080/10572252.2021.1930184.

Carey, Tamika L. 2014. "Take Your Place: Rhetorical Healing and Black Womanhood in Tyler Perry's Films." *Signs* 39, no. 4: 999–1021.

Carey, Tamika L. 2016. *Rhetorical Healing: The Reeducation of Contemporary Black Womanhood.* Albany: State University of New York Press.

Carliner, Saul. 2012. "The Three Approaches to Professionalization in Technical Communication." *Technical Communication* 59, no. 1 (February): 49–65. http://www.jstor.org/stable/43092920.

Carlson, Juliana, Cliff Leek, Erin Casey, Rich Tolman, and Christopher Allen. 2019. "What's in a Name? A Synthesis of 'Allyship' Elements from Academic and Activist Literature." *Journal of Family Violence* 35, no. 8 (July): 1–10.

Carrell, David. 1991. "Gender Scripts in Professional Writing Textbooks." *Journal of Business and Technical Communication* 5, no. 4 (October): 463–468. https://doi.org/10.1177/1050651991005004007.

Caswell, Nicole I., and Rebecca E. Johnson. 2022. "Centering Emotion at the Writing Center: An Approach to Tutor Training." In *Literacy and Learning in Times of Crisis: Emergent Teaching through Emergencies,* edited by Alice S. Horning, Sara P. Alvarez, Yana Kuchirko, Mark McBeth, Meghmala Tarafdar, and Missy Watson. New York: Peter Lang.

Castells, Manuel. 2012. *Networks of Outrage and Hope: Social Movements in the Internet Age.* Cambridge: Polity.

Center for Biological Diversity. 2020. "Dispersants." https://www.biologicaldiversity.org/programs/public_lands/energy/dirty_energy_development/oil_and_gas/gulf_oil_spill/dispersants.html.

Centers for Disease Control and Prevention. 2010. "Gulf Oil Spill Information for Pregnant Women." https://web.archive.org/web/20100624145144/https://www.bt.cdc.gov/gulfoilspill2010/2010gulfoilspill/pregnancy_oilspill.asp.

Chakravartty, Paula, and Yuezhi Zhao, eds. 2008. *Global Communications: Toward a Transcultural Political Economy.* Lanham, MD: Rowman and Littlefield.

Chandler, Gena E., and Jennifer Sano-Franchini. 2020. "Threat Assessment: Women of Color Teaching Ideological Critique in the Neoliberal Classroom." *Pedagogy* 20, no. 1: 87–100.

Christian, Barbara. 1987. "The Race for Theory." *Cultural Critique* 6: 51–63. https://doi.org/10.2307/1354255.

Cloud, Dana L. 2018. "Feminist Body Rhetoric in the #UnrulyMob, Texas, 2013." In *Unruly Rhetorics: Protest, Persuasion, and Publics,* edited by Jonathan Alexander, Susan C. Jarratt, and Nancy Welch, 27–44. Pittsburgh, PA: University of Pittsburgh Press.

Coletta, W. John. 1992. "The Ideologically Biased Use of Language in Scientific and Technical Writing." *Technical Communication Quarterly* 1, no. 1 (March): 59–70. https://doi.org/10.1080/10572259209359491.

Colton, Jared S., Steve Holmes, and Josephine Walwema. 2017. "From NoobGuides to #OpKKK: Ethics of Anonymous' Tactical Technical Communication." *Technical Communication Quarterly* 26, no. 1 (December): 59–75. https://doi.org/10.1080/10572252.2016.1257743.

Combahee River Collective. 1983. "The Combahee River Collective Statement." In *Home Girls: A Black Feminist Anthology,* edited by Barbara Smith, 264–274. New Brunswick, NJ: Rutgers University Press.

Connors, Robert. 1982. "The Rise of Technical Writing Instruction in America." *Journal of Technical Writing and Communication* 12, no. 4: 329–352.

Cook, Katsi. 2004. "Cycles of Continuous Creation." In *Ecological Medicine: Healing the Earth, Healing Ourselves,* edited by Kenny Ausubel, 155–164. San Francisco: Sierra Club Books.

Coole, Diana, and Samantha Frost. 2010. *New Materialisms: Ontology, Agency, and Politics.* Durham, NC: Duke University Press.

Coppola, Nancy W. 2012. "Professionalization of Technical Communication: Zeitgeist for Our Age, Introduction to This Special Issue (Part 2)." *Technical Communication* 59, no. 1 (February): 1–7. www.jstor.org/stable/43092916.

Corbett, Maryann Z. 1990. "Clearing the Air: Some Thoughts on Gender-Neutral Writing." *IEEE Transactions on Professional Communication* 33: 197–200.

Covi, Michelle P., Jennifer F. Brewer, and Donna J. Kain. 2021. "Sea Level Rise Hazardscapes of North Carolina: Perceptions of Risk and Prospects for Policy." *Ocean and Coastal Management* 212: 1–10.

Cox, Aimee Meredith, Tiye Giraud, Anna Gonzalez, Petra Kuppers, and Carrie Sandahl. 2008. "The Anarcha Project." *Liminalities: A Journal of Performance Studies* 4, no. 2. http://liminalities.net/42/anarcha/.

Cox, Matthew B. 2019. "Working Closets: Mapping Queer Professional Discourses and Why Professional Communication Studies Need Queer Rhetorics." *Journal of Business and Technical Communication* 33, no. 1 (September): 1–25. https://doi.org/10.1177/1050 651918798691.

Crasnow, Sharon, Alison Wylie, Wenda K. Bauchspies, and Elizabeth Potter. 2018. "Feminist Perspectives on Science." *Stanford Encyclopedia of Philosophy* (Spring). Edited by Edward N. Zalta. https://plato.stanford.edu/archives/spr2018/entries/feminist-science/.

Cray, Kate. 2021. "A High-Risk Group with a Tragically Low Vaccination Rate." *The Atlantic* (October). https://www.theatlantic.com/family/archive/2021/10/pregnant-people-low-vaccination-rate-covid-19/620458/.

Crenshaw, Kimberle W. 1991. "Mapping the Margins: Intersectionality, Identity, Politics, and Violence against Women of Color." *Stanford Law Review* 43, no. 6 (July): 1241–1299. https://doi.org/10.2307/1229039.

Cronn-Mills, Kirstin. 2000. "A Visible Ideology: A Document Series in a Women's Clothing Company." *Journal of Technical Writing and Communication* 30, no. 2 (April): 125–141. https://doi.org/10.2190/P8FR-R1D7-R4TW-6DE4.

Crowley, Sharon. 1998. *Composition in the University: Historical and Polemical Essays*. Pittsburgh, PA: University of Pittsburgh Press.

Crowley, Sharon, and Deborah Hawhee. 2012. *Ancient Rhetorics for Contemporary Students*, 5th ed. Boston: Longman.

Cuomo, Chris J. 1998. *Feminism and Ecological Communities: An Ethic of Flourishing*. New York: Routledge.

Dadurka, David, and Stacey Pigg. 2011. "Mapping Complex Terrains: Bridging Social Media and Community Literacies." *Community Literacy Journal* 6, no. 1 (Fall): 7–22. https://doi.org/10.1353/clj.2012.0010.

Daly, Herman E., and Joshua Farley. 2004. *Ecological Economics: Principles and Applications*. Washington, DC: Island Press.

Daugherty, Rachel Chapman. 2020. "Intersectional Politics of Representation: The Rhetoric of Archival Construction in Women's March Coalitional Memory." *Peitho* 22, no. 2 (Winter). https://cfshrc.org/article/intersectional-politics-of-representation-the-rhetoric-of-archival-construction-in-womens-march-coalitional-memory/.

Dauphin Island Real Estate, Inc. 2010. "Dauphin Island Real Estate Rentals." http://www.dauphinislandrentals.com/oil_spill.htm.

Dauphin Island Sea Lab. 2010. "Dauphin Island Sea Lab: Deepwater Horizon Oil Spill Response." http://oil.disl.org.

Davis, Carleigh J. 2018. "Memetic Rhetorical Theory in Technical Communication: Reconstructing Ethos in the Post-Fact Era." PhD dissertation, East Carolina University, Greenville, NC.

Davis, Marjorie T. 2001. "Shaping the Future of Our Profession." *Technical Communication* 48, no. 2: 139–144.

de Armas Ladd, Maria, and Marion Tangum. 1992. "What Difference Does Inherited Difference Make? Exploring Culture and Gender in Scientific and Technical Professions." *IEEE Transactions on Professional Communication* 35: 183–188.

de Certeau, Michel. 1984. *The Practice of Everyday Life*. Translated by Steven Rendall. Berkeley: University of California Press.

Deepwater Horizon Economic Claims Center. 2012. "Economic and Property Damage Claims." https://web.archive.org/web/20160303232902/http://www.deepwaterhorizoneconomicsettlement.com/index.php.

Deepwater Horizon Medical Benefits Claims Administrator. 2012. "Deepwater Horizon Medical Benefits." deepwaterhorizonmedicalsettlement.com.

De Hertogh, Lori Beth. 2018. "Feminist Digital Research Methodology for Rhetoricians of Health and Medicine." *Journal of Business and Technical Communication* 32, no. 4: 480–503.

Dell, Sherry A. 1990. "Promoting Equality of the Sexes through Technical Writing." *Technical Communication* 37, no. 3 (August): 248–251. https://www.jstor.org/stable/43094880.

Dell, Sherry A. 1992. "A Communication-Based Theory of the Glass Ceiling: Rhetorical Sensitivity and Upward Mobility within the Technical Organization." *IEEE Transactions on Professional Communication* 35, no. 4 (December): 230–235. https://doi.org/10.1109/47.180284.

de Onís, Kathleen M. 2015. "Lost in Translation: Challenging (White, Monolingual Feminism's) <Choice> with Justicia Reproductiva." *Women's Studies in Communication* 38, no. 1 (January): 1–19. https://doi.org/10.1080/07491409.2014.989462.

DePew, Kevin Eric, T. A. Fishman, Julia E. Romberger, and Bridget Fahey Ruetenik. 2006. "Designing Efficiencies: The Parallel Narratives of Distance Education and Composition Studies." *Computers and Composition* 23, no. 1 (January): 49–67. https://doi.org/10.1016/j.compcom.2005.12.005.

Derkatch, Colleen. 2016. *Bounding Biomedicine: Evidence and Rhetoric in the New Science of Alternative Medicine.* Chicago: University of Chicago Press.

DeTurk, Sara. 2011. "Allies in Action: The Communicative Experiences of People Who Challenge Social Injustice on Behalf of Others." *Communication Quarterly* 59, no. 5 (November–December): 569–590. https://doi.org/10.1080/01463373.2011.614209.

Ding, Huiling. 2009. "Rhetorics of Alternative Media in an Emerging Epidemic: SARS, Censorship, and Participatory Risk Communication." *Technical Communication Quarterly* 18, no. 4 (September): 327–350. https://doi.org/10.1080/10572250903149548.

Ding, Huiling. 2012. "Grassroots Emergency Health Risk Communication and Transmedia Public Participation: H1N1 Flu, Travelers from Epicenters, and Cyber Vigilantism." *Rhetoric, Professional Communication, and Globalization* 3: 15–35.

Ding, Huiling. 2013. "Transcultural Risk Communication and Viral Discourses: Grassroots Movements to Manage Global Risks of H1N1 Flu Pandemic." *Technical Communication Quarterly* 22, no. 1 (January): 126–149. https://doi.org/10.1080/10572252.2013.746628.

Dingo, Rebecca, and J. Blake Scott, eds. 2012. *The Megarhetorics of Global Development.* Pittsburgh, PA: University of Pittsburgh Press.

Dinshaw, Carolyn, Lee Edelman, Roderick A. Ferguson, Carla Freccero, Elizabeth Freeman, J. Halberstam, Annamarie Jagose, Christopher S. Nealon, and Tan Hoang Nguyen. 2007. "Theorizing Queer Temporalities: A Roundtable Discussion." *GLQ: A Journal of Lesbian and Gay Studies* 13, no. 2–3: 177–195. muse.jhu.edu/article/215002.

Dolmage, Jay Timothy. 2014. *Disability Rhetoric.* Syracuse, NY: Syracuse University Press.

Dombrowski, Paul M. 1994. "Challenger through the Eyes of Feyerabend." *Journal of Technical Writing and Composition* 24, no. 1: 7–18. https://doi.org/10.2190/WXNY-RQ8C-J9L8-0QEQ.

Dorpenyo, Isidore Kafui. 2019. "Risky Election, Vulnerable Technology: Localizing Biometric Use in Elections for the Sake of Justice." *Technical Communication Quarterly* 28, no. 4: 361–375.

Dragga, Sam. 1993. "Women and the Profession of Technical Writing: Social and Economic Influences and Implications." *Journal of Business and Technical Communication* 7, no. 3 (July): 312–321. https://doi.org/10.1177/1050651993007003002.

Dragga, Sam, and Dan Voss. 2001. "Cruel Pies: The Inhumanity of Technical Illustrations." *Technical Communication* 48, no. 3 (August): 265–274.

Durack, Katherine T. 1997. "Gender, Technology, and the History of Technical Communication." *Technical Communication Quarterly* 6, no. 3 (November): 249–260. https://doi.org/10.1207/s15427625tcq0603_2.

Eckholm, Erik, and Kim Severson. 2012. "Virginia Ultrasound Bill Passes as Other States Take Notice." *New York Times*, February 29. http://www.nytimes.com/2012/02/29/us/virginia-senate-passes-revised-ultrasound-bill.html.

Edbauer, Jenny. 2005. "Unframing Models of Public Distribution: From Rhetorical Situation to Rhetorical Ecologies." *Rhetoric Society Quarterly* 35, no. 4 (September): 5–24. https://doi.org/10.1080/02773940509391320.

Edenfield, Avery C., Steve Holmes, and Jared S. Colton. 2019. "Queering Tactical Technical Communication: DIY HRT." *Technical Communication Quarterly* 28, no. 3 (April): 177–191. https://doi.org/10.1080/10572252.2019.1607906.

Edwards, Jessica. 2018. "Race and the Workplace: Toward a Critically Conscious Pedagogy." In *Key Theoretical Frameworks: Teaching Technical Communication in the Twenty-First Century*, edited by Angela M. Haas and Michelle F. Eble, 268–286. Logan: Utah State University Press.

Edwards, Jessica, and Josie Walwema. 2022. "Black Women Imagining and Realizing Liberated Futures." *Technical Communication Quarterly* 3, no. 3: 245–262.

Ehrenreich, Barbara, and Deirdre English. 2010. *Witches, Midwives, and Nurses: A History of Women Healers*. New York: Feminist Press at the City University of New York.

Eiselt, Paula, and Tonya Lewis Lee, directors. 2022. *Aftershock*. Onyx Collective.

Enoch, Jessica. 2019. *Domestic Occupations: Spatial Rhetorics and Women's Work*. Carbondale: Southern Illinois University Press.

Eubanks, Virginia. 2011. *Digital Dead End: Fighting for Social Justice in the Information Age*. Cambridge, MA: MIT Press.

Evans, B. 2012. "Am I (Not) a Feminist Subject?" Paper presented at the Women's and Gender Studies Symposium, Illinois State University, Normal, spring.

Faber, Brenton, and Johndan Johnson-Eilola. 2002. "Migrations: Strategic Thinking about the Future(s) of Technical Communication." In *Reshaping Technical Communication: New Directions and Challenges for the Twenty-First Century*, edited by Barbara Mirel and Rachel Spilka, 135–148. New York: Lawrence Erlbaum.

Faris, Michael J. 2019. "Sex Education Comics: Feminist and Queer Approaches to Alternative Sex Education." *Journal of Multimodal Rhetorics* 3, no. 1. http://journalofmultimodalrhetorics.com/3-1-issue-faris.

Fernandez, Maria, Faith Wilding, and Michelle M. Wright. 2003. *Domain Errors! Cyberfeminist Practices: A subRosa Project*. New York: Autonomedia.

Fleckenstein, Kristie S., Clay Spinuzzi, Rebecca J. Rickly, and Carole Clark Papper. 2008. "The Importance of Harmony: An Ecological Metaphor for Writing Research." *College Composition and Communication* 60, no. 2 (December): 388–419.

Flower, Linda. 2002. "Intercultural Inquiry and the Transformation of Service." *College English* 65, no. 2 (November): 181–201. https://doi.org/10.2307/3250762.

Flynn, Elizabeth A. 1997. "Emergent Feminist Technical Communication." *Technical Communication Quarterly* 6, no. 3 (Summer): 313–320.

Flynn, Elizabeth A., Gerald Savage, Marsha Penti, Carol Brown, and Sarah Watke. 1991. "Gender and Modes of Collaboration in a Chemical Engineering Design Course." *Journal of Business and Technical Communication* 5, no. 4 (October): 444–462.

Flynn, Elizabeth A., Patricia Sotirin, and Ann Brady, eds. 2012. *Feminist Rhetorical Resilience*. Logan: Utah State University Press.

Flynn, John F. 1997. "Toward a Feminist Historiography of Technical Communication." *Technical Communication Quarterly* 6, no. 3 (Summer): 321–329.

Foss, Karen A., Sonja Foss, and Cindy Griffin. 1999. *Feminist Rhetorical Theories*. Thousand Oaks, CA: Sage.

Freedom University: About. 2022. https://www.freedom-university.org/home.

Freeman, Elizabeth. 2007. "Introduction." *GLQ: A Journal of Lesbian and Gay Studies* 13, no. 2–3 (June): 159–176. https://doi.org/10.1215/10642684-2006-029.
Freeman, Elizabeth. 2010. *Time Binds: Queer Temporalities, Queer Histories*. Durham, NC: Duke University Press.
Frieden, Betty. 1963. *The Feminine Mystique*. New York: Norton.
Frost, Erin A. 2013a. "Theorizing in Apparent Feminism in Technical Communication." PhD dissertation, Illinois State University, Normal.
Frost, Erin A. 2013b. "Transcultural Risk Communication on Dauphin Island: An Analysis of Ironically Located Responses to the Deepwater Horizon Disaster." *Technical Communication Quarterly* 22, no. 1: 50–66.
Frost, Erin A. 2016. "Apparent Feminism as a Methodology for Technical Communication and Rhetoric." *Journal of Business and Technical Communication* 30, no. 1 (September): 3–28. https://doi.org/10.1177/1050651915602295.
Frost, Erin A., and Michelle Eble. 2015. "Technical Rhetorics: Making Specialized Persuasion Apparent to Public Audiences." *Present Tense: A Journal of Rhetoric in Society* 2, no. 4 (February). https://www.presenttensejournal.org/volume-4/technical-rhetorics-making-specialized-persuasion-apparent-to-public-audiences/.
Frost, Erin A., and Michelle Eble. 2020. *Interrogating Gendered Pathologies*. Louisville, CO: Utah State University Press.
Frost, Erin A., Laura Gonzales, Marie Moeller, GPat Patterson, and Cecilia Shelton. 2021. "Unruly Bodies, Intersectionality, and Marginalization in Health and Medical Discourse." *Technical Communication Quarterly* 30, no. 3: 223–229. https://doi.org/10.1080/10572252.2021.1931457.
Frost, Erin A., and Angela Haas. 2017. "Seeing and Knowing the Womb: Examining Rhetorics of Fetal Ultrasound toward a Decolonization of Women's Bodies." *Computers and Composition* 43 (March): 88–105. https://doi.org/10.1016/j.compcom.2016.11.004.
Frost, Erin A., and Kellie Sharp-Hoskins. 2016. "Authorial Ethos as Location: How Technical Manuals Embody Authorial Ethos without Authors." In *Authorship Contested: Cultural Challenges to the Authentic, Autonomous Author*, edited by Amy Robillard and Ron Fortune, 71–88. New York: Routledge.
Gallop, Jane. 2002. *Anecdotal Theory*. Durham, NC: Duke University Press.
García Peña, Lorgia. 2022. *Community as Rebellion: A Syllabus for Surviving Academia as a Woman of Color*. Chicago: Haymarket.
Garrison-Joyner, Veronica, and Elizabeth Caravella. 2020. "Lapses in Literacy: Cultural Accessibility in Graphic Health Communication." *Technical Communication Quarterly* 29 (3): iii–xxv.
Gessner, David. 2015. "The Birds of British Petroleum." *Audubon Magazine* (July–August). https://www.audubon.org/magazine/july-august-2015/the-birds-british-petroleum.
GLAAD. 2020. "10 Ways to Be an Ally and a Friend." https://www.glaad.org/resources/ally/2.
Glenn, Cheryl. 1994. "Sex, Lies, and Manuscript: Refiguring Aspasia in the History of Rhetoric." *College Composition and Communication* 45, no. 2: 180–199. https://doi.org/10.2307/359005.
Glenn, Cheryl. 1997. *Rhetoric Retold: Regendering the Tradition from Antiquity to the Renaissance*. Carbondale: Southern Illinois University Press.
Glenn, Cheryl. 2018. *Rhetorical Feminism and This Thing Called Hope*. Carbondale: Southern Illinois University Press.
Goldenberg, Suzanne. 2010. "Deepwater Horizon Oil Spill: Tony Hayward's Worst Nightmare? Meet Wilma Subra, Activist Grandmother." *The Guardian*, June 20. https://amp.theguardian.com/environment/2010/jun/20/tony-hayward-bp-oil-spill.
Gonzales, Laura, and Angela Haas. 2019. "Building Familia from Scratch in/with TC: Accounting for Indigenous People, Paradigms, and Praxis." Paper presented at the

Association of Teachers of Technical Writing Convention, Pittsburgh, PA, March 12–13. https://attw.org/wp-content/uploads/2019/03/ATTW2019_WedSchedule-1.pdf.

Grabill, Jeffrey T. 2007. *Writing Community Change: Designing Technologies for Citizen Action*. Cresskill, NJ: Hampton.

Grabill, Jeffrey T., and W. Michele Simmons. 1998. "Toward a Critical Rhetoric of Risk Communication: Producing Citizens and the Role of Technical Communicators." *Technical Communication Quarterly* 7, no. 4 (Fall): 415–441. https://doi.org/10.1080/10572259809364640.

Green, McKinley. 2021. "Risking Disclosure: Unruly Rhetorics and Queer(ing) HIV Risk Communication on Grindr." *Technical Communication Quarterly* 30, no. 3: 271–284. https://doi.org/10.1080/10572252.2021.1930185.

Gregory, Rochelle. 2012. "A Womb with a View: Identifying the Culturally Iconic Fetal Image in Prenatal Ultrasound Provisions." *Present Tense: A Journal of Rhetoric in Society* 2, no. 2 (October). http://www.presenttensejournal.org/volume-2/a-womb-with-a-view-identifying-the-culturally-iconic-fetal-image-in-prenatal-ultrasound-provisions/.

Griffin, Susan. 2016. *Woman and Nature: The Roaring Inside Her*. Berkeley, CA: Counterpoint.

Grosz, Elizabeth. 2004. *The Nick of Time: Politics, Evolution, and the Untimely*. Durham, NC: Duke University Press.

Gurak, Laura J. 2018. "Ethos, Trust, and the Rhetoric of Digital Writing in Scientific and Technical Discourse." In *The Routledge Handbook of Digital Writing and Rhetoric*, edited by Jonathan Alexander and Jacqueline Rhodes, 124–131. New York: Taylor and Francis. https://doi.org/10.4324/9781315518497.

Gurak, Laura J., and Nancy L. Bayer. 1994. "Making Gender Visible: Extending Feminist Critiques of Technology to Technical Communication." *Technical Communication Quarterly* 3, no. 3 (June): 257–270. 10.1080/10572259409364571.

Guy-Sheftall, Beverly. 1995. *Words of Fire: An Anthology of African-American Feminist Thought*. New York: New York Press.

Haas, Angela M. 2007. "Wampum as Hypertext: An American Indian Intellectual Tradition of Multimedia Theory and Practice." *Studies in American Indian Literatures* 19, no. 4: 77–100. https://doi.org/10.1353/ail.2008.0005.

Haas, Angela M. 2008a. "A Rhetoric of Alliance: What American Indians Can Tell Us about Digital and Visual Rhetoric." PhD dissertation, Michigan State University, Lansing.

Haas, Angela M. 2008b. "Wired Wombs: A Rhetorical Analysis of Online Fertility Support Communities." In *Webbing Cyberfeminist Practice: Communities, Pedagogies, and Social Action*, edited by Kristine Blair, Radhika Gajjalaand, and Christine Tulley, 61–84. Creskill, NJ: Hampton.

Haas, Angela M. 2011. Graduate Course: Cultural Studies in Technical Communication. Illinois State University, Normal.

Haas, Angela M. 2012. "Race, Rhetoric, and Technology: A Case Study of Decolonial Technical Communication Theory, Methodology, and Pedagogy." *Journal of Business and Technical Communication* 26, no. 3 (April): 277–310. https://doi.org/10.1177/1050651912439539.

Haas, Angela M., and Michelle F. Eble. 2018. *Key Theoretical Frameworks: Teaching Technical Communication in the Twenty-First Century*. Louisville, CO: Utah State University Press.

Haas, Angela M., and Erin A. Frost. 2017. "Toward an Apparent Decolonial Feminist Rhetoric of Risk." In *Topic-Driven Environmental Rhetoric*, edited by Derek G. Ross, 168–186. New York: Routledge.

Haas, Angela M., Christine Tulley, and Kristine Blair. 2002. "Mentors versus Masters: Women's and Girls' Narratives of (Re)Negotiation in Web-Based Writing Spaces." *Computers and Composition* 19, no. 3 (October): 231–249. https://doi.org/10.1016/S8755-4615(02)00128-7.

Halberstam, Jack. 2002. "Queer Faces: Photography and Subcultural Lives." In *The Visual Culture Reader*, edited by Nicholas Mirzoeff, 96–108. London: Routledge.

Halberstam, Jack. 2005. *In a Queer Time and Place: Transgender Bodies, Subcultural Lives*. New York: New York University Press.

Hallenbeck, Sarah. 2012. "User Agency, Technical Communication, and the Nineteenth-Century Woman Bicyclist." *Technical Communication Quarterly* 21, no. 4 (August): 290–306.

Hamad, Ruby. 2020. *White Tears/Brown Scars: How White Feminism Betrays Women of Color*. Boca Raton, FL: Catapult.

Haraway, Donna J. 1989. *Primate Visions: Gender, Race, and Nature in the World of Modern Science*. New York: Routledge.

Haraway, Donna J. 1991. *Simians, Cyborgs, and Women: The Reinvention of Nature*. New York: Routledge.

Haraway, Donna J. 1996. *Modest Witness, Second Millennium: Female Man Meets Oncomouse*. New York: Routledge.

Harding, Sandra. 1986. *The Science Question in Feminism*. Ithaca, NY: Cornell University Press.

Harding, Sandra. 1992. "Rethinking Standpoint Epistemology: What Is 'Strong Objectivity?'" *Centennial Review* 36, no. 3: 437–470.

Harper, Kimberly C. 2017. "Black Lives Don't Matter Because Black Wombs Don't Matter: Exploring the Reproductive Rights of Black Mothers." In *Writing Networks for Social Justice*, edited by Don Unger and Liz Lane, 52–55. Constellations: A Cultural Rhetorics Publishing Space. http://constell8cr.com/4c4e/column_black_lives_dont_matter.

Harper, Kimberly C. 2020. *The Ethos of Black Motherhood in America: Only White Women Get Pregnant*. Lanham, MD: Lexington.

Harper, Kimberly C. 2021. "Tired as a Mutha: Black Mother Activists and the Fight for Affordable Housing and Health Care." *Technical Communication Quarterly* 30, no. 3: 230–240. https://doi.org/10.1080/10572252.2021.1930183.

Harper, Kimberly C. 2022. "Maternal Health Disparities." PBS North Carolina: NCImpact. https://video.pbsnc.org/video/maternal-health-disparities-haawyu/.

Hart-Davidson, William. 2001. "On Writing, Technical Communication, and Information Technology: The Core Competencies of Technical Communication." *Technical Communication* 48, no. 2 (May): 145–155. https://www.jstor.org/stable/43088814.

Hatcher, Alicia. 2021. "Performative Symbolic Resistance: Examining Symbolic Resistance Efforts of Black Professional Athletes through a New Methodological Analytical Framework." PhD dissertation, East Carolina University, Greenville, NC.

Hayles, N. Katherine. 1999. *How We Became Posthuman: Virtual Bodies in Cybernetics, Literature, and Informatics*. Chicago: University of Chicago Press.

Hayles, N. Katherine. 2019. "Can Computers Create Meanings? A Cyber/Bio/Semiotic Perspective." *Critical Inquiry* 46, no. 1 (Autumn): 32–55.

Hayles, N. Katherine. 2021. "Novel Corona: Posthuman Virus." *Critical Inquiry* 47, no. S2 (Winter): 68–72.

Haywood, Constance. 2019. "'I Do This for Us': Thinking through Reciprocity and Researcher-Community Relationships." *Gayle Morris Sweetland Digital Rhetoric Collaborative*. https://www.digitalrhetoriccollaborative.org/2019/12/10/i-do-this-for-us-thinking-through-reciprocity-researcher-community-relationships/.

Heaslip, Ashley M. 2014. "Non-Indigenous Ally." In *The SAGE Encyclopedia of Action Research*, edited by David Coghlan and Mary Brydon-Miller. Thousand Oaks, CA: Sage. https://ebookcentral.proquest.com/lib/eastcarolina/detail.action?docID=1712670.

Hemmings, Clare. 2011. *Why Stories Matter: The Political Grammar of Feminist Theory*. Durham, NC: Duke University Press.

Hesse-Biber, Sharlene N., and Patricia Leavy. 2007. *Feminist Research Practice: A Primer*. Thousand Oaks, CA: Sage.

Hill Collins, Patricia. 1990. *Black Feminist Thought: Knowledge, Consciousness, and the Politics of Empowerment*. New York: Routledge.

Holbrook, Sue Ellen. 1991. "Women's Work: The Feminizing of Composition." *Rhetoric Review* 9, no. 2 (May): 201–229. https://doi.org/10.1080/07350199109388929.
hooks, bell. 1989. *Talking Back: Thinking Feminist, Thinking Black.* Boston: South End.
hooks, bell. 1994. *Teaching to Transgress: Education as the Practice of Freedom.* New York: Routledge.
hooks, bell. 1995. *Art on My Mind: Visual Politics.* New York: New Press.
hooks, bell. 2004. "Black Vernacular: Architecture as Cultural Practice." In *Visual Rhetoric in a Digital World: A Critical Sourcebook,* edited by Carolyn Handa, 395–400. Boston: Bedford/St. Martin's.
Hull, Brittany, Cecilia D. Shelton, and Temptaous Mckoy. 2020. "Dressed but Not Tryin' to Impress: Black Women Deconstructing 'Professional' Dress." *Journal of Multimodal Rhetorics* 3, no. 2: 7–20.
Hurley, Elise Verzosa. 2018. "Spatial Orientations: Cultivating Critical Spatial Perspectives in Technical Communication Pedagogy." In *Key Theoretical Frameworks: Teaching Technical Communication in the Twenty-First Century,* edited by Angela M. Haas and Michelle F. Eble, 93–113. Logan: Utah State University Press.
Itchuaqiyaq, Cana Uluak. 2021. "MMU Scholar List." https://www.itchuaqiyaq.com/mmu-scholar-list.
Itchuaqiyaq, Cana Uluak, Avery C. Edenfield, and Keith Grant-Davie. 2021. "Sex Work and Professional Risk Communication: Keeping Safe on the Streets." *Journal of Business and Technical Communication* 36, no. 1. https://doi.org/10.1177/10506519211044190.
Itchuaqiyaq, Cana Uluak, and Breeanne Matheson. 2021. "Decolonial Dinners: Ethical Considerations of 'Decolonial' Metaphors in TPC." *Technical Communication Quarterly* 30, no. 3: 298–310. https://doi.org/10.1080/10572252.2021.1930180.
Jarratt, Susan C. 1998. *Rereading the Sophists: Classical Rhetoric Refigured.* Paperback ed. Carbondale: Southern Illinois University Press.
Jarratt, Susan C. 2002. "Sappho's Memory." *Rhetoric Society Quarterly* 32, no. 1 (Winter): 11–43. https://doi.org/10.1080/02773940209391219.
Jaworska, Sylvia. 2018. "Change but No Climate Change: Discourses of Climate Change in Corporate Social Responsibility Reporting in the Oil Industry." *International Journal of Business Communication* 55, no. 2 (February): 194–219. https://doi.org/10.1177/2329488417753951.
Jeyaraj, Joseph. 2004. "Liminality and Othering: The Issue of Rhetorical Authority in Technical Discourse." *Journal of Business and Technical Communication* 18, no. 1 (January): 9–38. https://doi.org/10.1177/1050651903257958.
Johnson, Maureen, Daisy Levy, Katie Manthey, and Maria Novotny. 2015. "Embodiment as an Emerging Key Concept in Feminist Rhetorics." *Peitho* 18, no. 1. https://cfshrc.org/wp-content/uploads/2015/10/18.1Johnsonetal.pdf.
Johnson, Paula. 2013. "His and Hers Healthcare." TED Video, 14:30. https://www.ted.com/talks/paula_johnson_his_and_hers_health_care?language=en.
Johnson, Robert R. 1998. *User-Centered Technology: A Rhetorical Theory for Computers and Other Mundane Artifacts.* Albany: State University of New York Press.
Johnson-Eilola, Johndan. 1996. "Relocating the Value of Work: Technical Communication in a Post-Industrial Age." *Technical Communication Quarterly* 5, no. 3 (November): 245–270. https://doi.org/10.1207/s15427625tcq0503_1.
Jones, Carolyn. 2012. " 'We Have No Choice': One Woman's Ordeal with Texas' New Sonogram Law." *Texas Observer,* March 15. https://www.texasobserver.org/we-have-no-choice-one-womans-ordeal-with-texas-new-sonogram-law/.
Jones, Natasha N. 2016a. "Narrative Inquiry in Human-Centered Design: Examining Silence and Voice to Promote Social Justice in Design Scenarios." *Journal of Technical Writing and Communication* 46, no. 4 (June): 471–492. https://doi.org/10.1177/0047281616653489.

Jones, Natasha N. 2016b. "The Technical Communicator as Advocate: Integrating a Social Justice Approach in Technical Communication." *Journal of Technical Writing and Communication* 46, no. 3 (April): 342–361. https://doi.org/10.1177/0047281616639472.

Jones, Natasha N. 2017. "Rhetorical Narratives of Black Entrepreneurs: The Business of Race, Agency, and Cultural Empowerment." *Journal of Business and Technical Communication* 31, no. 3: 319–349.

Jones, Natasha N., Laura Gonzales, and Angela M. Haas. 2021. "So You Think You're Ready to Build New Social Justice Initiatives? Intentional and Coalitional Pro-Black Programmatic and Organizational Leadership in Writing Studies." *WPA: Writing Program Administration* 44, no. 3: 29–35.

Jones, Natasha N., Kristen R. Moore, and Rebecca Walton. 2016. "Disrupting the Past to Disrupt the Future: An Antenarrative of Technical Communication." *Technical Communication Quarterly* 25, no. 4 (August): 211–229. https://doi.org/10.1080/10572252.2016.1224655.

Jones, Natasha N., and Miriam F. Williams. 2017. "The Social Justice Impact of Plain Language: A Critical Approach to Plain-Language Analysis." *IEEE Transactions on Professional Communication* 60, no. 4: 412–429. https://doi.org.10.1109/TPC.2017.2762964.

Jones, Natasha N., and Miriam F. Williams. 2018. "Technologies of Disenfranchisement: Literacy Tests and Black Voters in the US from 1890 to 1965." *Technical Communication* 65, no. 4: 371–386.

Juliette, Nakato. 2011. *Juliette's Story: A Jewelry Maker from Masese*. Produced by Amazima Ministries. September 8. YouTube video, 2:03. http://www.youtube.com/watch?v=g6Ww82OtQmU.

Katz, Steven B. 1992. "The Ethic of Expediency: Classical Rhetoric, Technology, and the Holocaust." *College English* 54, no. 3 (March): 255–275.

Keegan, Jon. 2019. "Blue Feed, Red Feed: See Liberal Facebook and Conservative Facebook, Side by Side." http://graphics.wsj.com/blue-feed-red-feed/.

Kerschbaum, Stephanie. 2014. *Toward a New Rhetoric of Difference*. Champaign, IL: National Council of Teachers of English.

Killingsworth, M. Jimmie. 2005. "From Environmental Rhetoric to Ecocomposition and Ecopoetics: Finding a Place for Professional Communication." *Technical Communication Quarterly* 14, no. 4 (November): 359–373. https://doi.org/10.1207/s15427625tcq1404_1.

Kimme Hea, Amy C., Adrienne Crump, Elise Verzosa, Crystal N. Fodrey, Anita Furtner Archer, Jennifer Haley-Brown, Ashley J. Holmes, Marissa M. Juarez, Londie T. Martin, and Jenna Vinson. 2012. "Space | Event | Movement: Reflections on a Spatial and Visual Rhetorics Graduate Course." *Kairos* 16, no. 3. https://kairos.technorhetoric.net/16.3/praxis/hea-et-al/index.html.

King, Bradley S., and John D. Gibbins. 2011. *Health Hazard Evaluation of Deepwater Horizon Response Workers: Health Hazard Evaluation Report, HETA 2010-0115 and 2010-0129-3138*. Springfield, VA: National Institute for Occupational Safety and Health.

King, Thomas. 2008. *The Truth about Stories: A Native Narrative*. Minneapolis: University of Minnesota Press.

Knoblauch, A. Abby, and Marie E. Moeller. 2022. *Bodies of Knowledge: Embodied Rhetorics in Theory and Practice*. Louisville, CO: Utah State University Press.

Koerber, Amy. 2013. *Breast or Bottle? Contemporary Controversies in Infant-Feeding Policy and Practice*. Columbia: University of South Carolina Press.

Koerber, Amy. 2018. *From Hysteria to Hormones: A Rhetorical History*. University Park: Pennsylvania State University Press.

Koerber, Amy. 2002. "Postmodernism, Resistance, and Cyberspace: Making Rhetorical Spaces for Feminist Mothers on the Web." *Women's Studies in Communication* 24: 218–240.

Kolasinski, Nicholas J. 2017. "The Relationship between Ally Identity and Implicit Associations towards Lesbian, Gay, Bisexual, Transgender, and Queer (LGBTQ) Individuals." PhD dissertation, William James College, Newton, MA.

Koonce, Danielle. 2021. "Rural Resistance and the Environment: Understanding How Rural Black Communities Engage in Environmental Justice." October 28. Emerging Scholars Symposium Research/Clinical Presentations, East Carolina University Office for Equity and Diversity, Greenville, NC.

Kopelson, Karen. 2003. "Rhetoric on the Edge of Cunning; or, the Performance of Neutrality (Re)considered as a Composition Pedagogy for Student Resistance." *College Composition and Communication* 55, no. 1 (September): 115–146. https://doi.org/10.2307/3594203.

Kopelson, Karen. 2020. "Rhetoric on the Edge of Cunning Revisited: Of Truth and Lies in an Extra Urgent Sense." *Pedagogy* 20, no. 1 (January): 13–20. https://doi.org/10.1215/15314200-7878953.

Kostelnick, Charles. 2007. "The Rhetorical Minefield of Risk Communication." *Journal of Business and Technical Communication* 21, no. 1 (January): 21–22. https://doi.org/10.1177/1050651906293508.

Kruschek, Gina. 2019. " 'You Have Herpes, Now What': Stigma in Healthcare Systems and Disclosure Rhetorics." PhD dissertation, East Carolina University, Greenville, NC.

Kynell, Teresa. 1999. "Technical Communication from 1850–1950: Where Have We Been?" *Technical Communication Quarterly* 8: 143–151.

Kynell-Hunt, Teresa, and Gerald Savage. 2003. *Power and Legitimacy in Technical Communication: The Historical and Contemporary Struggle for Professional Status*. Amityville, NY: Baywood.

LaDuc, Linda, and Amanda Goldrick-Jones. 1994. "The Critical Eye, the Gendered Lens, and 'Situated' Insights—Feminist Contributions to Professional Communication." *Technical Communication Quarterly* 3, no. 3 (March): 245–256. https://doi.org/10.1080/10572259409364570.

LaDuke, Winona. 1999. *All Our Relations: Native Struggles for Land and Life*. Cambridge, MA: South End.

LaDuke, Winona. 2005. *Recovering the Sacred: The Power of Naming and Claiming*. Cambridge, MA: South End.

Lay, Mary M. 1989. "Interpersonal Conflict in Collaborative Writing: What We Can Learn from Gender Studies." *Journal of Business and Technical Communication* 3, no. 2 (September): 5–28. https://doi.org/10.1177/105065198900300202.

Lay, Mary M. 1991. "Feminist Theory and the Redefinition of Technical Communication." *Journal of Business and Technical Communication* 5: 348–370.

Lay, Mary M. 1993. "Gender Studies: Implications for the Professional Communication Classroom." In *Professional Communication: The Social Perspective*, edited by Nancy Roundy Blyler and Charlotte Thralls, 215–229. Newbury Park, CA: Sage.

Lay, Mary M., Janice J. Monk, and Deborah Rosenfelt, eds. 2001. *Encompassing Gender: Integrating International Studies and Women's Studies*. New York: Feminist Press at City University of New York.

Ledbetter, Lehua. 2018. "The Rhetorical Work of YouTube's Beauty Community: Relationship- and Identity-Building in User-Created Procedural Discourse." *Technical Communication Quarterly* 27, no. 4 (October): 287–299.

Lippincott, Gail. 2003. "Rhetorical Chemistry: Negotiating Gendered Audiences in Nineteenth-Century Nutrition Studies." *Journal of Business and Technical Communication* 17, no. 1 (January): 10–49. https://doi.org/10.1177/1050651902238544.

Lithwick, Dahlia. 2012. "Virginia's Proposed Ultrasound Law Is an Abomination." *Slate Magazine*, February 16. http://www.slate.com/articles/double_x/doublex/2012/02/virginia_ultrasound_law_women_who_want_an_abortion_will_be_forcibly_penetrated_for_no_medical_reason.html.

Lockett, Alexandria. 2019. "Scaling Black Feminisms: A Critical Discussion about the Digital Labor of Representation." In *Humans at Work in the Digital Age: Forms of Digital Textual Labor*, edited by Andrew Pilsch and Shawna Ross, 250–266. New York: Routledge.

Logan, Shirley W. 1999. *"We Are Coming": The Persuasive Discourse of Nineteenth Century Black Women.* Carbondale: Southern Illinois University Press.

Loupe, Diane. 2010. "Oil Spill's Harm to Pregnant Women Unknown." *Women's E-News: Covering Women's Issues, Changing Women's Lives,* July 22. https://womensenews.org/2010/07/oil-spills-harm-pregnant-women-unknown/.

Love, Heather. 2007. *Feeling Backward: Loss and the Politics of Queer History.* Cambridge, MA: Harvard University Press.

Lundgren, Regina E. 1994. *Risk Communication: A Handbook for Communicating Environmental, Safety, and Health Risks.* Columbus, OH: Battelle.

Lundgren, Regina E., and Andrea H. McMakin. 2009. *Risk Communication: A Handbook for Communicating Environmental, Safety, and Health Risks,* 3rd ed. Hoboken, NJ: Institute of Electrical and Electronics Engineers.

Lunsford, Andrea, ed. 1995. *Reclaiming Rhetorica.* Pittsburgh, PA: University of Pittsburgh Press.

MacArthur Foundation. 2005. "Wilma Alpha Subra." https://www.macfound.org/fellows/625/.

Mallette, Jennifer C. 2017. "Writing and Women's Retention in Engineering." *Journal of Business and Technical Communication* 31, no. 4 (October): 417–442. http://dx.doi.org/10.1177/1050651917713253.

Marback, Richard. 2004. "The Rhetorical Space of Robben Island." *Rhetoric Society Quarterly* 34, no. 2 (June): 7–27. https://doi.org/10.1080/02773940409391279.

Martin, Emily. 1991. "The Egg and the Sperm: How Science Has Constructed a Romance Based on Stereotypical Male-Female Roles." *Signs: Journal of Women in Culture and Society* 16, no. 3: 485–501.

Martin, Emily. 2001. *The Woman in the Body: A Cultural Analysis of Reproduction.* Boston: Beacon.

McCallum, E. L., and Mikko Tuhkanen, eds. 2011. *Queer Times, Queer Becomings.* New York: State University of New York Press.

McDowell, Earl E. 2003. "Tracing the History of Technical Communication from 1850–2000: Plus a Series of Survey Studies." Washington, DC: US Department of Education, Educational Resources Information Center.

McElmurry, Shawn P. 2016. "Rapid Response to Contaminants in Flint Drinking Water." Washington, DC: National Institutes of Health, National Institute of Environmental Health Sciences. Project #1R21ES027199–01. https://grantome.com/grant/NIH/R21-ES027199-01.

McIlwain, Charlton D. 2019. *Black Software: The Internet and Racial Justice, from the AfroNet to Black Lives Matter.* New York: Oxford University Press.

Mckoy, Temptaous. 2019. "Y'all Call It Technical and Professional Communication, We Call It #ForTheCulture: The Use of Amplification Rhetorics in Black Communities and Their Implications for Technical and Professional Communication Studies." PhD dissertation, East Carolina University, Greenville, NC.

Mckoy, Temptaous, Cecilia D. Shelton, Carleigh Davis, and Erin A. Frost. 2022. "Embodying Public Feminisms: Collaborative Intersectional Models for Engagement." *IEEE Transactions on Professional Communication* 65, no. 1: 70–86.

Mckoy, Temptaous, Cecilia D. Shelton, Donnie Sackey, Natasha N. Jones, Constance Haywood, Ja'La Wourman, and Kimberly C. Harper. 2020. *CCCC Black Technical and Professional Communication Position Statement with Resource Guide.* https://cccc.ncte.org/cccc/black-technical-professional-communication.

Mckoy, Temptaous, Cecilia D. Shelton, Donnie Sackey, Natasha N. Jones, Constance Haywood, Ja'La Wourman, and Kimberly C. Harper. 2022. "Introduction to Special Issue: Black Technical and Professional Communication." *Technical Communication Quarterly* 31, no. 3: 221–228.

McRuer, Robert. 2006. *Crip Theory: Cultural Signs of Queerness and Disability*. New York: New York University Press.

Meagher, Sharon M., and Patrice DiQuinzio, eds. 2005. *Women and Children First: Feminism, Rhetoric, and Public Policy*. New York: State University of New York Press.

Medina-López, Kelly. 2018. "Rasquache Rhetorics: A Cultural Rhetorics Sensibility." *Constellations: A Cultural Rhetorics Publishing Space* 1, no. 1: 1–20.

Miller, Carolyn R. 1989. "What's Practical about Technical Writing?" In *Technical Writing: Theory and Practice*, edited by Bertie E. Fearing and W. Keats Sparrow, 14–24. New York: Modern Language Association.

Moeller, Marie E., and Erin A. Frost. 2016. "Food Fights: Cookbook Rhetorics, Monolithic Constructions of Womanhood, and Field Narratives in Technical Communication." *Technical Communication Quarterly* 25, no. 1: 1–11. https://doi.org/10.1080/10572252.2016.1113025.

Mohanty, Chandra Talpade. 1988. "Under Western Eyes: Feminist Scholarship and Colonial Discourses." *Feminist Review* 30, no. 1 (Autumn): 61–88.

Mohanty, Chandra Talpade. 2003. "'Under Western Eyes' Revisited: Feminist Solidarity through Anticapitalist Struggles." *Signs: Journal of Women in Culture and Society* 28, no. 2 (Winter): 499–535.

Moreno, Renee M. 2002. "'The Politics of Location': Text as Opposition." *College Composition and Communication* 54, no. 2 (December): 222–242.

Morris, Kerri K. 2020. "Women and Bladder Cancer: Listening Rhetorically to Healthcare Disparities." In *Interrogating Gendered Pathologies*, edited by Erin A. Frost and Michelle Eble, 142–151. Louisville, CO: Utah State University Press.

Mountford, Roxanne. 2001. "On Gender and Rhetorical Space." *Rhetoric Society Quarterly* 31, no. 1 (Winter): 41–71. https://doi.org/10.1080/02773940109391194.

Mullin, Rick. 2020. "Wilma Subra: An Unstoppable Pioneer in Environmental Chemistry and Community Advocacy." *Chemistry and Engineering News*, January 17. https://cen.acs.org/people/awards/Wilma-Subra-unstoppable-pioneer-environmental/98/i3.

Muñoz, José Esteban. 2009. *Cruising Utopia: The Then and There of Queer Futurity*. New York: New York University Press.

Munster, Anna. 2006. *Materializing New Media: Embodiment in Information Aesthetics*. Lebanon, NH: University Press of New England.

Murphy, Mollie K. 2017. "What's in the World Is in the Womb: Converging Environmental and Reproductive Justice through Synecdoche." *Women's Studies in Communication* 40, no. 2 (April): 155–171. https://doi.org/10.1080/07491409.2017.1285839.

Nakamura, Lisa. 2008. *Digitizing Race: Visual Cultures of the Internet*. Minneapolis: University of Minnesota Press.

National Research Council. 1989. *Improving Risk Communication*. Washington, DC: National Academy Press.

Neeley, Kathryn A. 1992. "Woman as Mediatrix: Women as Writers on Science and Technology in the Eighteenth and Nineteenth Centuries." *IEEE Transactions on Professional Communication* 35, no. 4 (December): 208–216. https://doi.10.1109/47.180281.

Nelms, R. Gerald. 2004. "The Rise of Technical Writing Instruction in America." In *Central Works in Technical Communication*, edited by Johndan Johnson-Eilola and Stuart Selber, 3–19. New York: Oxford.

Nixon, Rob. 2011a. "Slow Violence." *The Chronicle*. https://www.chronicle.com/article/Slow-Violence/127968.

Nixon, Rob. 2011b. *Slow Violence and the Environmentalism of the Poor*. Cambridge, MA: Harvard University Press.

Noble, Safiya. 2018. *Algorithms of Oppression: How Search Engines Reinforce Racism*. New York: New York University Press.

Novotny, Maria. 2015. "reVITALize Gynecology: Reimagining Apparent Feminism's Methodology in Participatory Health Intervention Projects." *Communication Design Quarterly* 3, no. 4 (September): 61–74. https://doi.org/10.1145/2826972.2826978.

Novotny, Maria, and Les Hutchinson. 2019. "Data Our Bodies Tell: Towards Critical Feminist Action in Fertility and Period Tracking Applications." *Technical Communication Quarterly* 28, no. 4 (April): 332–360. https://doi.org/10.1080/10572252.2019.1607907.

Obergefell v. Hodges. 135 S. Ct. 2584 (2015).

Occupational Safety and Health Administration. 2012. "OSHA Fact Sheet: Carbon Monoxide Poisoning." https://www.osha.gov/OshDoc/data_General_Facts/carbonmonoxide-factsheet.pdf.

Ogle, Robbin S., and Susan Jacobs. 2002. *Self-Defense and Battered Women Who Kill: A New Framework.* Westport, CT: Praeger.

Olinger, Andrea. 2021. "Closing Remarks." Watson Conference, Louisville, KY, April 23.

Ornatowski, Cezar M. 1992. "Between Efficiency and Politics: Rhetoric and Ethics in Technical Writing." *Technical Communication Quarterly* 1, no. 1 (March): 91–103. https://doi.org/10.1080/10572259209359493.

Ott, Riki. 2010. "Bio-Remediation or Bio-Hazard? Dispersants, Bacteria, and Illness in the Gulf." *HuffPost.* https://www.huffpost.com/entry/bio-remediation-or-bio-ha_b_720461.

Owens, Kim Hensley. 2015. *Writing Childbirth: Women's Rhetorical Agency in Labor and Online.* Carbondale: Southern Illinois University Press.

Owens, Kim Hensley, and Catherine Molloy. 2022. "Looking for a Mind [and Body and Heart] at Work." *Rhetoric of Health and Medicine* 5, no. 3: 241–249.

People v. Aris. 1989. 215 Cal. App. 3d 1178, 264 Cal. Rptr. 167, 1989 Cal. App. LEXIS 1187 (Court of Appeal of California, Fourth Appellate District, Division Two, November 17, 1989).

People v. Beasley. 1993. 251 Ill. App. 3d 872, 622 N.E.2d 1236, 1993 Ill. App. LEXIS 1649, 190 Ill. Dec. 919 (Appellate Court of Illinois, Fifth District, November 1, 1993, Filed).

Petersen, Emily January. 2014. "Redefining the Workplace: The Professionalization of Motherhood through Blogging." *Journal of Technical Writing and Communication* 44, no. 3 (September): 277–296. https://doi.org/10.2190/TW.44.3.d.

Petersen, Emily January. 2019. "The 'Reasonably Bright Girls': Accessing Agency in the Technical Communication Workplace through Interactional Power." *Technical Communication Quarterly* 28, no. 1 (November): 21–38. https://doi.org/10.1080/10572252.2018.1540724.

Petersen, Emily January, and Ryan M. Moeller. 2016. "Using Antenarrative to Uncover Systems of Power in Mid-Twentieth Century Policies on Marriage and Maternity at IBM." *Journal of Technical Writing and Communication* 46, no. 3 (March): 362–386. https://doi.org/10.1177/0047281616639473.

Petersen, Emily January, and Rebecca Walton. 2018. "Bridging Analysis and Action: How Feminist Scholarship Can Inform the Social Justice Turn." *Journal of Business and Technical Communication* 32, no. 4 (June): 416–446. https://doi.org/10.1177/1050651918780192.

Plough, Alonzo, and Sheldon Krimsky. 1990. "The Emergence of Risk Communication Studies: Social and Political Context." In *Readings in Risk*, edited by Theodore S. Glickman and Michael Gough, 223–230. Washington, DC: Resources for the Future.

Potts, Liza. 2009. "Using Actor Network Theory to Trace and Improve Multimodal Communication Design." *Technical Communication Quarterly* 18, no. 3 (June): 281–301. https://doi.org/10.1080/10572250902941812.

Potts, Liza. 2014. *Social Media in Disaster Response: How Experience Architects Can Build for Participation.* New York: Routledge.

Pouncil, Floyd, and Nick Sanders. 2022. "The Work Before: A Model for Coalitional Alliance toward Black Futures in Technical Communication." *Technical Communication Quarterly* 31, no. 3: 283–297.

Pratt, Mary Louise. 1999. "Arts of the Contact Zone." In *Ways of Reading: An Anthology for Writers*, edited by David Bartholomae and Anthony Petroksky, 33–40. New York: Bedford/St. Martin's.

Raign, Kathryn R. 2018. "Finding Our Missing Pieces—Women Technical Writers in Ancient Mesopotamia." *Journal of Technical Writing and Communication* 49, no. 3 (September): 338–364. https://doi.org/10.1177/0047281618793406.

Ramazanoğlu, Carolina, and Janet Holland. 2009. *Feminist Methodology: Challenges and Choices*. London: Sage.

Ratcliffe, Krista. 1995. *Anglo American Feminist Challenges to the Rhetorical Tradition*. Carbondale: Southern Illinois University Press.

Ratcliffe, Krista. 2005. *Rhetorical Listening: Identification, Gender, Whiteness*. Carbondale: Southern Illinois University Press.

Ratcliffe, Krista. 2018. "Afterword." In *Composing Feminist Interventions: Activism, Engagement, Praxis*, edited by Kristine L. Blair and Lee Nickoson, 505–510. Fort Collins, CO: WAC Clearinghouse.

Rauch, Susan. 2012. "The Accreditation of Hildegard von Bingen as Medieval Female Technical Writer." *Journal of Technical Writing and Communication* 42, no. 4 (September): 393–411. https://doi.org/10.2190/TW.42.4.d.

Reuters Staff. 2015. "BP Settles Oil Spill–Related Claims with Halliburton, Transocean." https://www.reuters.com/article/us-halliburton-bp-oilspill-idUSKBN0O52LL 20150521.

Reynolds, Nedra. 1998. "Composition's Imagined Geographies: The Politics of Space in the Frontier, City, and Cyberspace." *College Composition and Communication* 50, no. 1 (September): 12–35.

Rhodes, Jacqueline. 2018. "Slutwalk Is Not Enough: Notes toward a Critical Feminist Rhetoric." In *Unruly Rhetorics: Protest, Persuasion, and Publics*, edited by Jonathan Alexander, Susan C. Jarratt, and Nancy Welch, 88–104. Pittsburgh, PA: University of Pittsburgh Press.

Richards, Daniel P. 2017. "Reconstituting Causality: Accident Reports as Posthuman Documentation." In *Topic-Driven Environmental Rhetoric*, edited by Derek G. Ross, 149–167. New York: Routledge.

Ridolfo, Jim, and Dànielle Nicole DeVoss. 2009. "Composing for Recomposition: Rhetorical Velocity and Delivery." *Kairos* 13, no. 2 (January). http://www.technorhetoric.net/13.2/topoi/ridolfo_devoss/intro.html.

Ridolfo, Jim, and William Hart-Davidson. 2019. *Rhet Ops: Rhetoric and Information Warfare*. Pittsburgh, PA: University of Pittsburgh Press.

Rifkind, Lawrence J., and Loretta F. Harper. 1992. "Cross-Gender Immediacy Behaviors and Sexual Harassment in the Workplace: A Communication Paradox." *IEEE Transactions on Professional Communication* 35, no. 4 (December): 236–241.

Robinson, Joy, Lisa Dusenberry, Liz Hutter, Halcyon Lawrence, Andy Frazee, and Rebecca Burnett. 2019. "State of the Field: Teaching with Digital Tools in the Writing and Communication Classroom." *Computers and Composition* 54. https://doi.org/10.1016/j.compcom.2019.102511.

Robvais, Raquel M. 2020. "We Are No Longer Invisible." *Poroi* 15, no. 1: 1–18.

Rohrer-Vanzo, Valentina, Tobias Stern, Elisabeth Ponocny-Seliger, and Peter Schwarzbauer. 2016. "Technical Communication in Assembly Instructions: An Empirical Study to Bridge the Gap between Theoretical Gender Differences and Their Practical Influence." *Journal of Business and Technical Communication* 30, no. 1 (September): 29–58. https://doi.org/10.1177/1050651915602292.

Ross, Derek G., ed. 2017. *Topic-Driven Environmental Rhetoric*. New York: Routledge.

Ross, Susan Mallon. 1994. "A Feminist Perspective on Technical Communicative Action: Exploring How Alternative Worldviews Affect Environmental Remediation Efforts." *Technical Communication Quarterly* 3, no. 3 (Summer): 325–342.

Rothschild, Joan A. 1981. "A Feminist Perspective on Technology and the Future." *Women's Studies International Quarterly* 4, no. 1: 65–74. https://doi.org/10.1016/S0148-0685(81)96373-9.

Roundtree, Aimee. 2017. "Social Health Content and Activity on Facebook: A Survey Study." *Journal of Technical Writing and Communication* 47, no. 3: 300–329.

Royal, Cindy. 2005. "A Meta-Analysis of Journal Articles Intersecting Issues of Internet and Gender." *Journal of Technical Writing and Communication* 35, no. 4 (October): 403–429.

Royster, Jacqueline Jones, and Gesa E. Kirsch. 2012. *Feminist Rhetorical Practices: New Horizons for Rhetoric, Composition, and Literacy Studies*. Carbondale: Southern Illinois University Press.

Rude, Carolyn. 1979. "A Humanistic Rationale for Technical Communication." *College English* 40, no. 6: 610–617.

Ruiz, Iris D. 2018. "La indigena: Risky Identity Politics and Decolonial Agency as Indigenous Consciousness." *Journal of Pan African Studies* 11, no. 6. https://link.gale.com/apps/doc/A541103915/AONE?u=googlescholar&sid=bookmark-AONE&xid=f12e3d3b.

Rushe, Dominic. 2012. "BP Sues Halliburton for Deepwater Horizon Oil Spill Clean-up Costs: Oil Group BP Lays Blame for Deepwater Disaster on Haliburton's Cement Work and Seeks Unspecified Damages." *The Guardian*, January 3. http://www.guardian.co.uk/business/2012/jan/03/bp-sues-halliburton-over-deepwater.

Sackey, Donnie Johnson. 2020. "One-Size-Fits-None: A Heuristic for Proactive Value Sensitive Environmental Design." *Technical Communication Quarterly* 29, no. 1: 33–48. https://doi.org/10.1080/10572252.2019.1634767.

Sánchez, Fernando. 2019. "Trans Students' Right to Their Own Gender in Professional Communication Courses: A Textbook Analysis of Attire and Voice Standards in Oral Presentations." *Journal of Technical Writing and Communication* 49, no. 2 (December): 183–212. https://doi.org/10.1177/0047281618817349.

Sandman, Peter M. 2020. "The Peter M. Sandman Risk Communication Website." http://www.psandman.com.

Sandoval, Chela. 2000. *Methodology of the Oppressed*. Minneapolis: University of Minnesota Press.

Sauer, Beverly A. 1992. "The Engineer as Rational Man: The Problem of Imminent Danger in a Non-Rational Environment." *IEEE Transactions on Professional Communication* 35, no. 4 (December): 242–249. https://doi.10.1109/47.180286.

Sauer, Beverly A. 1993. "Sense and Sensibility in Technical Documentation: How Feminist Interpretation Strategies Can Save Lives in the Nation's Mines." *Journal of Business and Technical Communication* 7, no. 1 (January): 63–83. https://doi.org/10.1177/1050651993007001004.

Sauer, Beverly A. 1994. "Sexual Dynamics of the Profession: Articulating the Ecriture Masculine of Science and Technology." *Technical Communication Quarterly* 3, no. 3: 309–323. https://doi.org/10.1080/10572259409364574.

Sauer, Beverly A. 2003. *The Rhetoric of Risk: Technical Documentation in Hazardous Environments*. Mahwah, NJ: Lawrence Erlbaum Associates.

Sauer, Beverly A. 2010. "Engineering Safety: Lessons in Risk Communication from the BP Disaster." GSFC Systems Engineering Seminar Series, NASA–Goddard Space Flight Center. http://ses.gsfc.nasa.gov/ses_data_2010/101102_Sauer_Abstract.htm.

Savage, Gerald J. 1996. "Redefining the Responsibilities of Teachers and the Social Position of the Technical Communicators." *Technical Communication Quarterly* 5, no. 3 (November): 309–327.

Savage, Gerald J. 1999. "The Process and Prospects for Professionalizing Technical Communication." *Journal of Technical Writing and Communication* 29, no. 4 (October): 355–381.

Savage, Gerald J. 2003. "Tricksters, Fools, and Sophists: Technical Communication as Postmodern Rhetoric." In *Power and Legitimacy in Technical Communication: Strategies for*

Professional Status, vol. 2, edited by Teresa Kynell-Hunt and Gerald J. Savage, 167–193. Amityville, NY: Baywood.

Savage, Gerald J. 2010. "Program Assessment, Strategic Modernism, and Professionalization Politics: Complicating Coppola and Elliot's 'Relational Model.' " In *Assessment in Technical and Professional Communication*, edited by Margaret N. Hundleby and Jo Allen, 161–168. Amityville, NY: Baywood.

Savage, Gerald, and Kyle Mattson. 2011. "Perceptions of Racial and Ethnic Diversity in Technical Communication Programs." *Programmatic Perspectives* 3, no. 1: 5–57.

Savage, Gerald, and Natalia Matveeva. 2011. "Toward Racial and Ethnic Diversity in Technical Communication Programs: A Study of Technical Communication in Historically Black Colleges and Universities and Tribal Colleges and Universities in the United States." *Programmatic Perspectives* 3, no. 1: 152–179.

SB 484 2012. "Abortion; Ultrasound Required at Least 24 Hours Prior to Undergoing Procedure." http://leg1.state.va.us/cgi-bin/legp504.exe?121+sum+SB484.

Schell, Eileen E. 1998. *Gypsy Academics and Mother-Teachers: Gender, Contingent Labor, and Writing Instruction*. Portsmouth, NH: Boynton.

Schuster, Mary Lay. 2015. "My Career and the 'Rhetoric of' Technical Writing and Communication." *Journal of Technical Writing and Communication* 45, no. 4 (May): 381–391. https://doi.org/10.1177/0047281615585754.

Scott, J. Blake. 2004. *Risky Rhetoric: AIDS and the Cultural Practices of HIV Testing*. Carbondale: Southern Illinois University Press.

Scott, J. Blake, Bernadette Longo, and Katherine V. Wills. 2006. *Critical Power Tools: Technical Communication and Cultural Studies*. Albany: State University of New York Press.

Serano, Julia. 2013. *Excluded: Making Feminist and Queer Movements More Inclusive*. Berkeley: Seal Press.

Sexsmith, Sinclair. 2012. "Jack Halberstam: Queers Create Better Models of Success." *Lamba Literary*, February. https://lambdaliterary.org/2012/02/jack-halberstam-queers-create-better-models-of-success/.

Sharp-Hoskins, Kellie C. 2012. "What Counts? Who Counts? A Methodology for Leveraging Perspective on the Terministic Management of Language and Bodies." PhD dissertation, Illinois State University, Normal.

Sharpe, Christina. 2016. *In the Wake: On Blackness and Being*. Durham, NC: Duke University Press.

Shelton, Cecilia. 2019a. "On Edge: A Techné of Marginality." PhD dissertation, East Carolina University, Greenville, NC.

Shelton, Cecilia. 2019b. "Shifting Out of Neutral: Centering Difference, Bias, and Social Justice in a Business Writing Course." *Technical Communication Quarterly* 29, no. 1 (July): 18–32. https://doi.org/10.1080/10572252.2019.1640287.

Simmons, W. Michele, and Jeffrey T. Grabill. 2007. "Toward a Civic Rhetoric for Technologically and Scientifically Complex Places: Invention, Performance, and Participation." *College Composition and Communication* 58, no. 3 (February): 419–448. https://www.jstor.org/stable/20456953.

Simmons, W. Michele, and Meredith W. Zoetewey. 2012. "Productive Usability: Fostering Civic Engagement and Creating More Useful Online Spaces for Public Deliberation." *Technical Communication Quarterly* 21, no. 3 (March): 251–276. https://doi.org/10.1080/10572252.2012.673953.

Skinner, Carolyn. 2012. "Incompatible Rhetorical Expectations: Julia W. Carpenter's Medical Society Papers, 1895–1899." *Technical Communication Quarterly* 21, no. 4 (April): 307–324. https://doi.org/10.1080/10572252.2012.686847.

Smith, Allegra W. 2014. "Porn Architecture: User Tagging and Filtering in Two Online Porn Communities." *Communication Design Quarterly* 3, no. 1: 17–22.

Smith, Andrea. 2015. *Conquest: Sexual Violence and American Indian Genocide*. Durham, NC: Duke University Press.

Solinger, Rickie. 2001. *Beggars and Choosers: How the Politics of Choice Shapes Adoption, Abortion, and Welfare in the United States.* New York: Hill and Wang.

Sovereign Nation of the Chitimacha. 2015. "Tribal History." In *Chitimacha Tribe of Louisiana.* http://www.chitimacha.gov/history-culture/tribal-history.

Sperandio, Elena. 2015. "Influences of Technical Documentation and Its Translation on Efficiency and Customer Satisfaction." In *Communication Practices in Engineering, Manufacturing, and Research for Food and Water Safety,* edited by David Wright, 145–170. IEEE PCS Professional Engineering Communication Series. Hoboken, NJ: Wiley.

Stanford University Medical Center. 2006. "Transgender Experience Led Stanford Scientist to Critique Gender Difference." https://www.sciencedaily.com/releases/2006/07/060714174545.htm.

Starke-Meyerring, Doreen, and Melanie Wilson. 2008. "Learning Environments for a Globally Networked World: Emerging Visions." In *Designing Globally Networked Learning Environments: Visionary Partnerships, Policies, and Pedagogies,* edited by Doreen Starke-Meyerring and Melanie Wilson, 1–17. Rotterdam, Netherlands: Sense Publishers.

State v. Norman. 1989. 324 N.C. 253, 378 S.E.2d 8, 1989 N.C. LEXIS 158 (Supreme Court of North Carolina, April 5, 1989, Filed).

State v. Stewart. 1988. 243 Kan. 639, 763 P.2d 572, 1988 Kan. LEXIS 186 (Supreme Court of Kansas, October 21, 1988, Opinion Filed).

Stephens, Sonia H., and Daniel P. Richards. 2020. "Story Mapping and Sea Level Rise: Listening to Global Risks at Street Level." *Communication Design Quarterly* 8, no. 1 (May): 5–18. https://doi.org/10.1145/3375134.3375135.

Stratman, James F. 2007. "A Reflection on Risk Communication, Metacommunication, and Rhetorical Stases in the Aspen-EPA Superfund Controversy." *Journal of Business and Technical Communication* 21, no. 1 (January): 23–26. https://doi.org/10.1177/1050651906293523.

Sullivan, Patricia, and Kristen Moore. 2013. "Time Talk: On Small Changes That Enact Infrastructural Mentoring for Undergraduate Women in Technical Fields." *Journal of Technical Writing and Communication* 43, no. 3 (July): 333–354. https://doi.org/10.2190/TW.43.3.f.

Sun, Huatong. 2009. "Toward a Rhetoric of Locale: Localizing Mobile Messaging Technology into Everyday Life." *Journal of Technical Writing and Communication* 39, no. 3 (June): 245–261. https://doi.org/10.2190/TW.39.3.c.

Swacha, Kathryn Yankura. 2018. "'Bridging the Gap between Food Pantries and the Kitchen Table': Teaching Embodied Literacy in the Technical Communication Classroom." *Technical Communication Quarterly* 27, no. 3 (June): 261–282. https://doi.org/10.1080/10572252.2018.1476589.

Tebeaux, Elizabeth. 1993. "Technical Writing for Women of the English Renaissance." *Written Communication* 10, no. 2 (April): 164–199. https://doi.org/10.1177/0741088393010002002.

Tebeaux, Elizabeth, and Mary M. Lay. 1992. "Images of Women in Technical Books from the English Renaissance." *IEEE Transactions on Professional Communication* 35, no. 4 (December): 196–207. https://doi.org/10.1109/47.180280.

Teston, Christa. 2012. "Moving from Artifact to Action: A Grounded Investigation of Visual Displays of Evidence during Medical Deliberations." *Technical Communication Quarterly* 21, no. 3 (January): 187–209. https://doi.org/10.1080/10572252.2012.650621.

Teston, Christa, and S. Scott Graham. 2012. "Stasis Theory and Meaningful Public Participation in Pharmaceutical Policy." *Present Tense: A Journal of Rhetoric in Society* 2, no. 2 (October). http://www.presenttensejournal.org/volume-2/stasis-theory-and-meaningful-public-participation-in-pharmaceutical-policy-making/.

Texas Health and Safety Code. 2012. "Woman's Right to Know Act." 171.002–064.

Thierry, Mike. 2011. "Charter Fishing with Capt. Mike out of Dauphin Island, AL." http://captainmikeonline.com.

Thompson, Isabelle. 1999. "Women and Feminism in Technical Communication: A Qualitative Content Analysis of Journal Articles Published in 1989 through 1997." *Journal of Business and Technical Communication* 13, no. 2 (April): 154–178. https://doi.org/10.1177/1050651999013002002.

Thompson, Isabelle, and Elizabeth Overman Smith. 2006. "Women and Feminism in Technical Communication—an Update." *Journal of Technical Writing and Communication* 36, no. 2 (April): 183–199. https://doi.org/10.2190/4JUC-8RAC-73H6-N57U.

Thralls, Charlotte, and Nancy Roundy Blyler. 2002. "Cultural Studies: An Orientation for Research in Professional Communication." In *Research in Technical Communication*, edited by Laura J. Gurak and Mary M. Lay, 185–209. Westport, CT: Praeger.

Tillery, Denise. 2018. *Commonplaces of Scientific Evidence in Environmental Discourses*. New York: Routledge.

Town of Dauphin Island. 2010. "Oil Spill Preparations." May. http://www.townofdauphinisland.org/default.asp?id=119.

Unger, Nancy C. 2010. "From Jook Joints to Sisterspace: The Role of Nature in Lesbian Alternative Environments in the United States." In *Queer Ecologies: Sex, Nature, Politics, Desire*, edited by Catriona Mortimer-Sandilands and Bruce Erickson, 173–198. Bloomington: Indiana University Press.

van Slyck, Phyllis. 1997. "Repositioning Ourselves in the Contact Zone." *College English* 59, no. 2 (February): 149–170.

Vatz, Richard E. 1968. "The Myth of the Rhetorical Situation." *Philosophy and Rhetoric* 6, no. 3 (Summer): 154–161.

Vaughn, Jeannette. 1989. "Sexist Language—Still Flourishing." *Technical Writing Teacher* 16, no. 1 (Winter): 33–40.

Vealey, Kyle P., and Alex Layne. 2018. "Of Complexity and Caution: Feminism, Object-Oriented Ontology, and the Practices of Scholarly Work." In *Feminist Rhetorical Science Studies: Human Bodies, Posthumanist Worlds*, edited by Amanda K. Booher and Julie Jung, 50–83. Carbondale: Southern Illinois University Press.

Walker, Alice. 1983. *In Search of Our Mothers' Gardens: Womanist Prose*. New York: Harcourt.

Walls, Douglas. 2011. "Review of Digital Dead End: Fighting for Social Justice in the Information Age by Virginia Eubanks." *Community Literacy Journal* 6, no. 1 (Fall): 97–100.

Walton, Rebecca, Kristen R. Moore, and Natasha N. Jones. 2019. *Technical Communication after the Social Justice Turn: Building Coalitions for Action*. New York: Routledge.

Walton, Rebecca, Maggie Zraly, and Jean Pierre Mugengana. 2015. "Values and Validity: Navigating Messiness in a Community-Based Research Project in Rwanda." *Technical Communication Quarterly* 24, no. 1 (October): 45–69. https://doi.org/10.1080/10572252.2015.975962.

Walwema, Josephine, and Felicita Arzu Carmichael. 2021. "'Are You Authorized to Work in the U.S.?' Investigating 'Inclusive' Practices in Rhetoric and Technical Communication Job Descriptions." *Technical Communication Quarterly* 30, no. 2: 107–122. https://doi.org/10.1080/10572252.2020.1829072.

Wang, Hua. 2021. "Chinese Women's Reproductive Justice and Social Media." *Technical Communication Quarterly* 30, no. 3: 285–297. https://doi.org/10.1080/10572252.2021.1930178.

Washington, Harriet A. 2006. *Medical Apartheid: The Dark History of Medical Experimentation on Black Americans from Colonial Times to the Present*. New York: Harlem Moon.

Welch, Kathleen Ethel. 2005. "Technical Communication and Physical Location: Topoi and Architecture in Computer Classrooms." *Technical Communication Quarterly* 14, no. 3 (November): 335–344. https://doi.org/10.1207/s15427625tcq1403_12.

Welsh, Deborah M. 2014. "'Surprise! You're Dead': The Deepwater Horizon Disaster and Opening Statements in the Court of Public Opinion." PhD dissertation, East Carolina University, Greenville, NC.

White, Kate, Suzanne Kesler Rumsey, and Stevens Amidon. 2016. "Are We 'There' Yet? The Treatment of Gender and Feminism in Technical, Business, and Workplace Writing Studies." *Journal of Technical Writing and Communication* 46, no. 1 (September): 27–58.

Wildcat, Daniel R. 2009. *Red Alert! Saving the Planet with Indigenous Knowledge*. Golden, CO: Fulcrum.

Williams, Miriam F., and Octavio Pimentel. 2014. *Communicating Race, Ethnicity, and Identity in Technical Communication*. Amityville, NY: Baywood. https://doi.org/10.4324/9781315232584.

Williams, Miriam F., and Octavio Pimentel, eds. 2016. *Communicating Race, Ethnicity, and Identity in Technical Communication*. New York: Routledge.

Williams, Patricia J. 1992. *The Alchemy of Race and Rights: Diary of a Law Professor*. Cambridge, MA: Harvard University Press.

Woods, Clyde. 2010. *In the Wake of Hurricane Katrina: New Paradigms and Social Visions*. Baltimore, MD: Johns Hopkins University Press.

Woody, Cassandra. 2020. "Re-engaging Rhetorical Education through Procedural Feminism: Designing First-Year Writing Curricula That Listen." *College Composition and Communication* 71, no. 3: 481–507.

Wray, Amanda, and Elise Verzosa Hurley. 2016. "Feminist Rhetorical Praxis: Everyday Feminism as Public Agora." *Res Rhetorica* 3, no. 2: 37–51. https://doi.10.17380/rr2016.2.1.

Wright, David, ed. 2015. *Communication Practices in Engineering, Manufacturing, and Research for Food and Water Safety*. IEEE PCS Professional Engineering Communication Series. Hoboken, NJ: Wiley.

Yusuf, Modupe, and Veena Namboodri Schioppa. 2022. "A Technical Hair Piece: Metis, Social Justice, and Technical Communication in Black Hair Care on YouTube." *Technical Communication Quarterly* 31, no. 3: 263–282.

Youngblood, Susan A. 2012. "Balancing the Rhetorical Tension between Right-to-Know and Security in Risk Communication: Ambiguity and Avoidance." *Journal of Business and Technical Communication* 26, no. 1 (November): 33–62. https://doi.org/10.1177/1050651911421123.

Zachry, Mark, and Charlotte Thralls, eds. 2007. *Communicative Practices in Workplaces and the Professions: Cultural Perspectives on the Regulation of Discourse and Organizations*. Amityville, NY: Baywood.

Zurn, Perry, Danielle S. Bassett, and Nicole C. Rust. 2020. "The Citation Diversity Statement: A Practice of Transparency, a Way of Life." *Trends in Cognitive Sciences* 24, no. 9: 669–672. https://doi.org/10.1016/j.tics.2020.06.009.

INDEX

access, 43–44
accomplice, 47–48
affirmative technical communication, 28
Aftershock (documentary), 7
agency, 36, 71, 75–76, 142, 146–147
Alabama Department of Public Health, 112
Ali, Nujood, 15
allies, 31, 39, 46–50, 66, 143
anecdote, 44
Anglo-Persian Oil Company (APOC), 123
Anonymous (hacktivist group), 26
antenarrative (as methodology), 27
anti-choice legislation, 33, 36
antitoxics movement, 125–126
apparency, 4–6, 13–16, 40, 43–44, 127–129, 133
apparent feminist methodology, definition of, 31; purpose of, 38; necessity of, 39
Arab Spring, 130
Aspasia, 13–14, 35
Association of Teachers of Technical Writing (ATTW) 25, 39, 144
Atlanta Washerwomen's strike, 28
Atlanta, Georgia, 146
audience, diversity of, 12, 14, 17, 31, 54–55, 66, 128
auto-colonization, 21

Basquiat, Jean-Michel, 60
Battered Woman Syndrome, 72
bias, 38–39, 41–42, 67, 107. *See also* objectivity
biosemiotics, 64
Black technical and professional communication (BTPC), 16
body, 78, 80, 87. *See also* embodiment
British Petroleum (BP), characteristics of, 90, 120; documents authored by, 94–95, 108, 115, 121; history of, 104, 123–124; public opinion of, 123; relationship to locals following DHD, 108–110, 113; sponsorship, 100
Bureau of Safety and Environmental Enforcement (BSEE), 103
Bureau of Safety and Environmental Enforcement Well Control Rule (2010), 103

Carpenter, Julia W., 30
Cartesian split, 78
Cavarero, Adriana, 26
Center for Biological Diversity (CBD), 90
Challenger (disaster), 10
childbirth, 7
Chitimacha Tribe, 124
chrononormativity, 83
citizen communication, 106, 111
Civil Rights movement, 19
climate change, 84
coalition, 7, 41, 46, 61, 144
Combahee River Collective, 134
community (as in being in community with), 144–145, 147
community inquiry, 120
Conference on College Composition and Communication (CCCC), 16
Cook, Katsi, 59
Corexit®, 89–90. *See also* dispersant
Council for Programs in Technical and Scientific Communication (CPTSC), 25
Covid-19, 6, 44, 84, 140
crisis, parameters of the term, 74, 77, 84, 136–137, 141. *See also* slow crisis
crisis communication, 3, 34, 133
crisis rhetoric, 136
critical race theory, 47, 73, 83, 129
cultural studies, 10, 14, 20, 23–24, 26, 46, 150n12
culture, definition of, 120

Dauphin Island Real Estate, Inc., 94
Dauphin Island Sea Lab (DISL), 4, 94, 99, 111
Deepwater Horizon Claims Center, 102, 110
Deepwater Horizon Disaster (DHD), discussion of human health missing from, 88–89; as exigence for theorizing apparent feminisms, 36, 85; histories of, 123–124; long-term consequences of, 92, 118
Deepwater Horizon Disaster Settlement, 9
Democrat (party affiliation), 137

178 INDEX

digital rhetorics, 14, 49, 69, 94, 131–133
dispersant, 89–91, 102–104

East Carolina University, 18, 85, 153n8
eco-criticism, 70
Economic and Property Damage site, 102
EcoTour (multimedia environmental advocacy project in a state park), 130
ectogenesis, 62
efficiency: reimagined as tenet of feminist apparency, 66, 68, 83, 88, 117, 133–135, 143–147; standard definition of, 103–104
efficiency models, as applied in DHD rhetorics, 110–112, 114
efficiency rhetorics, 54, 77, 106, 134
embodiment, 28–29, 39, 58, 79–80, 100; in digital spaces, 129–133
Enheduanna, 26
environmental disaster, 7, 76, 110, 133
environmental health, 9, 97–98, 124
environmental justice, 5, 59, 125–126, 147
Environmental Protection Agency (US), 22, 59, 89
ethics, 51
exigency, 8. *See also* urgency
expertise, 16–17, 50, 118, 149–150n5
Exxon Valdez spill, 90

Feminisms: history in TPC, 10, 20–26, 31–32; plurality of, 3, 31, 57, 117, 138
#feministarmy, 137
feminist rhetorics, 46, 57–65
Flint, Michigan, 5
Freedom University, 144–145

General Motors, 59
global warming. *See* climate change
god term, 3, 58, 125, 139
Grindr, 8
guerrilla media, 69, 106, 115–117
Gulf of Mexico, 85, 123
gynotechnics, 60

Halliburton (corporation), 89, 152n2
happiness (theorized), 82, 137
Harvard School of Medicine, 7
hazardous material (HAZMAT), 113
health (human): absent concerns about, 94, 114, 123–125; apparency of human health crises, 140–142; as contributor to efficiency, 88–89. *See also* health communication
health communication, 3, 5–9
health risks, 8–9, 92, 133
historiography, 27, 28, 60, 82, 142
hog farming, 124

Houghton, Michigan, 129
Human Resources Department (Mobile County Department of Health), 96
Hurricane Katrina, 109

IEEE Transactions on Professional Communication, 21
Illinois State University, 18, 30, 41
in vitro fertilization (IVF), 62
inclusion (politics of), 64
intersectional feminism, 16, 18, 26, 30–31, 105–106. *See also* feminist rhetorics

Journal of Business and Technical Communication (JBTC), 20
journalist, research role, 9, 97. *See also* reporter
Juliette, Nakato, 15

legal rhetorics, 9, 36, 73
location, theorized, 69, 119. *See also* space
Louisiana Environmental Action Network, 126

MacArthur grant, 126
Macondo area, 89, 104, 123–124
maleness, valuing of in technical communication, 22
master narratives, 118
materialism, 28, 86–87
medicine, 5
Middle East (world region), 123
Minerals Management Service (MMS), 123
Mississippi Canyon Block 252, 124
Mobile County Health Department (MCHD), 96, 114, 139
Mohawk community, 22, 59
Mothers' Milk Project, 59
Mount Vernon, Alabama, 99

narrative (as methodology), 27, 58
National Audubon Society, 4
National Institutes of Health, 5
Natural Resources Defense Council, 103
Nazi, 54
NCImpact, 7
neutrality, 48, 135. *See also* objectivity
New England Kitchen (the), 24
Nightingale, Florence, 24
non-feminist, 41, 46, 48, 50, 66
North Carolina Agricultural and Technical State University, 7
North Carolina State University, 39

Obergefell v. Hodges (legal case, 2015), 79
objectivity, 31, 38, 41–42, 71, 82

Occupational Safety and Health Administration, 91
Occupy Wall Street, 130
oral rhetorics, 109
orientations, 36, 63, 85, 145
origin stories, 12–13, 17, 128, 142
Other (as concept), 61, 78, 88

Patrick, Dan (Senator, R-Houston), 38
Pelican Reef restaurant (Theodore, Alabama), 95, 99, 109
People v. Aris (legal case, 1989), 73
People v. Beasley (legal case, 1993), 73
postcolonialism, 47, 63, 83
post-feminism, 45, 67, 143
posthumanism, 61, 64, 151*n*10
posttraumatic stress, 72
procedural feminism, 65
professional communication, 107
public intellectual, 40, 58

queer temporality, 34, 78–82
queer theory, 28, 63, 79, 82

reporter, research role, 44, 96, 99–100, 152*n*2. *See also* journalist
reproductive justice, 7, 36, 66
Republican (party affiliation), 153*n*6
resilience, 122, 128, 143, 146
reVITALize Gynecology infertility initiative, 29
rhetorical distancing, 84
rhetorical feminisms, 32–33
rhetorical situation, 42, 48, 53, 111, 131–133
Richards, Ellen Swallow, 24
risk communication, 10, 68–70, 87, 109, 116, 121–122
risk, construction of 6, 69, 110, 113, 121
Roe v. Wade (legal case, 1973), 151*n*2

Sappho, 13–14
self-determination, 28
self-erasure, 48
self-identification, 138, 151*n*5
sickle cell disease (SCD), 127
slow crisis, definition of, 71–75
slow death, 75–77
slow homicide, 73–74
slow violence, 75
social framework, 50, 74
social media, 5, 44, 130. *See also* digital rhetorics

Socrates, 35
space, as theoretical concept, 63–66, 119
stakeholder input, 29
State v. Norman (legal case, 1989), 73
State v. Stewart (legal case, 1988), 73
Stewart, Mike, 72–73
Stewart, Peggy, 72–73
Subra, Wilma Alpha, 126–127
Subra Company (Louisiana), 126
subversive historiography, 142. *See also* historiography
Supreme Court (US), 79
sustainability, 55, 70, 143
Syracuse University (NY), 145

tar balls, 89–90
Technical Communication Quarterly (TCQ), 8, 16, 22, 30, 142
terministic screens, 55, 68. *See also* slow crisis
Texas City Refinery, 124
Texas Health and Safety Code, 38, 66
Theodore, Alabama, 95, 99
third world women, 17, 61
transcultural analysis, 105, 117–118
transcultural communication, 106, 107, 115, 116, 117–120
Transocean Ltd., 89, 123

University of Louisiana at Lafayette, 126
urgency, 8–9, 42, 44, 45, 71–72, 77. *See also* slow crisis
US Army Corps of Engineers, 114
US Coast Guard, 89

visuality, 128–129
von Bingen, Hildegard, 13

Walker, Lenore, 72. *See also* Battered Woman Syndrome
Water Resources Center (East Carolina University), 85
Weeksville Heritage Center (Brooklyn), 7
Weems, Carrie Mae, 60
Western (part of the world), 17, 61–63
wildlife, cleanup efforts, 90, concern for, 109, 112
Woman's Right to Know Act (2003; amended 2012), 38, 42, 46, 51
womanism, 16, 138
Women's E-News, 102
Women's March, 134

ABOUT THE AUTHOR

Erin A. Clark (previously Erin A. Frost) is a technical communication and rhetoric scholar at East Carolina University. Her work combines technical communication and gender studies, and this confluence often happens in the realms of environmental studies, rhetorics of risk, and rhetorics of health and medicine.

www.ingramcontent.com/pod-product-compliance
Lightning Source LLC
Chambersburg PA
CBHW060603080526
44585CB00013B/675